Technological Innovation:
Government / Industry Cooperation

Technological Innovation: Government / Industry Cooperation

edited by
ARTHUR GERSTENFELD
Worcester Polytechnic Institute

with the assistance of
ROBERT BRAINARD
National Science Foundation

A WILEY – INTERSCIENCE PUBLICATION
JOHN WILEY & SONS
New York / Chichester / Brisbane / Toronto

Copyright © 1979 by John Wiley & Sons, Inc.

All rights reserved. Published simultaneously in Canada.

Reproduction or translation of any part of this work beyond that permitted by Sections 107 or 108 of the 1976 United States Copyright Act without the permission of the copyright owner is unlawful. Requests for permission or further information should be addressed to the Permissions Department, John Wiley & Sons, Inc.

Library of Congress Cataloging in Publication Data:

Main entry under title:
Technological innovation–government/industry
   cooperation.

     "A Wiley–Interscience publication.'
   Based on papers presented at a conference held in
Geneva, June 1977.
    Includes bibliographical references and index.
     1. Technological innovations—Congresses.
I. Gerstenfeld, Arthur, 1927–
T173.8.P8   338.4'7'6   78-14800
ISBN 0-471-03647-1   '

Printed in the United States of America

10 9 8 7 6 5 4 3 2 1

Contributors to the Conference

The following were contributors to the five day conference held in Geneva, Switzerland in June, 1977, on which this volume is based.

Robert Brainard, *National Science Foundaton, Washington, D.C.*

Klaus Brockhoff, *University of Kiel, Kiel, West Germany*

Umberto Colombo, *Montedison, Milano, Italy*

Joan G. Cox, *University of Essex, Essex, England*

Steven Dollond, *Arthur D. Little Company, London, England*

Herbert I. Fusfeld, *Kennecott Copper Corporation, New York, New York*[*]

Arthur Gerstenfeld, *Worcester Polytechnic Institute, Worcester, Massachusetts*

Michael Gibbons, *University of Manchester, Manchester, England*

Earl H. Hess, *Lancaster Laboratories, Inc., Lancaster, Pennsylvania*

Hiroshi Inose, *University of Tokyo, Tokyo, Japan*

Aubrey Kagan, *Laboratory for Clinical Stress Research, Stockholm, Sweden*

Regina M. Kelly, *Department of Commerce, Washington, D.C.*

George Kozmetsky, *University of Texas, Austin, Texas*

Helmar Krupp, *Institut für Systemtechnik und Innovations, Karlsruhe, West Germany*

Leonard L. Lederman, *National Science Foundation, Washington, D.C.*

[*] Current affiliation, Director, Center for Science and Technology Policy, New York University, New York.

Edwin Mansfield, *University of Pennsylvania, Philadelphia, Pennsylvania*

Gerhardt Mensch, *International Institute of Management, Berlin, West Germany*

Howard K. Nason, *I.R.I. Research Corporation, St. Louis, Missouri*

Richard R. Nelson, *Yale University, New Haven, Connecticut*

A. E. Pannenborg, *Phillips Industries, Eindhoven, The Netherlands*

K. L. R. Pavitt, *University of Sussex, Brighton, England*

Peter R. Payne, *Payne, Inc., Annapolis, Maryland*

Roy Rothwell, *University of Sussex, Sussex, England*

Albert H. Rubenstein, *Northwestern University, Evanston, Illinois*

D. Sahal, *Portland State University, Portland, Oregon*

R. Schultz, *E.I.R.M.A., Paris, France*

Achim A. Stoehr, *WFG, Frankfurt, West Germany*

Karl A. Stroetmann, *ABT. Associates GMBH, Heidelberg, West Germany*

H. Thiemann, *Nestle Headquarters, Vevey, Switzerland*

Michiyuki Uenohara, *Nippon Electric Company, Japan*

Walter E. L. Zegveld, *Organization for Industrial Research T.N.O., Deflt, The Netherlands*

Acknowledgments

Through the support of the National Science Foundation (NSF) and the Division of International Programs a group of distinguished persons from nine different countries, representing many different organizations from industry, government, and the academic community, were brought together by NSF for a meeting held in Geneva, Switzerland in June, 1977. Particular appreciation is expressed to Robert Brainard of the National Science Foundation for complete support from the conception of the ideas within this book through to completion. For many months prior to the meeting, contributors pondered and wrote on various specific aspects that fell within the broad purview of "Government/Industry Cooperation for Technological Innovation." Upon completion of the meeting, the papers presented and discussed there were revised and edited. They are provided in what we hope is an important and interesting contribution.

The editor wishes to express his deep appreciation for the wonderful cooperation of each of the participants and for the editing assistance of Lynne Marcus and the secretarial and administrative support of Mary Eaton. I recognize from the outset that the problems associated with the ways in which industry and government can cooperate to nurture technological innovation could not be "solved." Rather, they involve vitally important issues that I believe deserve careful and thoughtful focus. The papers and discussions on which the chapters in this book are based have the objective not only of focusing on the issues and raising vitally important questions, but also of including specific options. I feel I have raised many important questions that should be asked, and I have suggested many options. There is a long way to go, however, and I hope that readers will raise more questions and search for more solutions while putting into effect policies and practices that are ready for implementation.

A.G.
R.B.

Contents

1 Technological Innovation: Some Introductory Remarks   1
   The Editor

PART ONE
OVERVIEW: ECONOMIC AND SOCIAL ASPECTS OF INNOVATION

2 Present and Future Context for Innovation   7
   K. L. R. Pavitt

3 Returns from Industrial Innovation   18
   Edwin Mansfield

4 The Effects of Technological Change on Employment   20
   Roy Rothwell

5 Technological Innovation and International Trade Patterns   41
   Regina M. Kelly

6 Human and Social Aspects of Technological Change   57
   Aubrey Kagan

PART TWO
INDUSTRY: POLICIES AND PRACTICES FOR INDUSTRIAL INNOVATION

7 The Environment for Industrial Innovation in the United States   69
   Howard K. Nason

8	New Technology-Based Firms Arthur D. Little, Ltd.	80
9	Innovation in Medium and Small Industrial Firms Karl A. Stroetmann	93
10	The Small, High-Technology Firm Peter R. Payne	104
11	The Small, High-Technology Professional Service Firm Earl H. Hess	111

PART THREE

GOVERNMENT: POLICIES AND PRACTICES FOR INDUSTRIAL INNOVATION

Country Reports

12	Government Policy and Innovation in England Michael Gibbons and P. J. Gummett	121
13	Government Policy and Innovation in Japan Hiroshi Inose	140
14	Government Policy and Innovation in the United States Leonard L. Lederman	159

Special Topics

15	Government Action and the Innovation Process Albert H. Rubenstein	165
16	Government Regulations and Innovation Arthur Gerstenfeld	176
17	Improving the Management of Technology Klaus Brockhoff	186
18	Closer Integration of Science Policy and Economic Policy Richard R. Nelson	196

PART FOUR
GOVERNMENT-INDUSTRY COOPERATION

19	Innovation Strategies for Government and Industry Umberto Colombo	209
20	Aligning Industry Planning and Government Policy A. E. Pannenborg	227
21	National Science Policy and the Private Sector Herbert I. Fusfeld	234
22	Technology and Psychology: A Query Gerhardt Mensch	242
23	Summary	245

INDEX 271

**Technological Innovation:
Government / Industry
Cooperation**

1 Technological Innovation: Some Introductory Remarks

The Editor

This volume considers the subject of technological innovation focusing on industry and government cooperation as a guide and stimulus toward beneficial technological innovation. It follows what we believe is a logical sequence by first examining the economic and social aspect of innovation. This leads to the second part that focuses on industry's role in technological innovation and a third part that focuses on the government's role in technological innovation. The final section blends the previous parts of the book and considers how industry and government can cooperate for technological innovation.

Chapter 2 examines the present and future context for innovation and suggests that innovation in the future will take a different direction and be directed toward better utilization of resources. This, of course, is already happening with increased concern for gas mileage and home insulation. These issues were of far less concern to industry or government officials a few years ago.

The third chapter in Part 1 focuses on the issue of who the major beneficiary from industrial innovation is. This chapter describes a study showing that the public or the user generally receives a greater benefit from innovation than innovating firms. This conclusion is based on consideration of the high risks and the low probabiliites of success for the company that invests in innovations. This implication is important because it can help explain why the United States has reached recently a technological plateau. For, indeed if noninnovation investments for a firm offer a higher rate of return than innovation (when all factors are considered), then firms may indeed choose to invest less in innovation activities. It is perhaps through industry-government cooperation that this situation can be ameliorated.

2 Technological Innovation: Some Introductory Remarks

Chapter 4 examines the effects of technical change on employment. Case studies and analyzed data indicate that companies using less technological input are less successful than those using the newest technologies. This may sound like a tautology, but this phenomenon is not limited to higher-technology industries but may indeed be of even more importance to the basic industries that still form the backbone of most countries' economies. We are lead to conclude that no industries are free of the spectre of technological change.

Chapter 5 shows that the United States exports a far higher percentage of higher-technology products than other countries. However, some European countries use high-technology products mainly in their domestic markets, and the percentage of exports for these higher-technology products is no higher than in less technically developed countries.

Technological change can have considerable benefits on worker environment, the subject of Chapter 6. Several examples from Sweden are presented, especially that of the Volvo factory, which is described as a plant that blends new technologies with the social psychology of work groups. This model is presented to the readers as an alternative to the more traditional production techniques.

Part 2, which focuses on the industry component of the government-industry relationship, starts by considering the environment for industrial innovation in the United States. This is followed by a contrasting view from Europe with particular emphasis on new ventures in technology-based firms. The final three chapters in Part 2 consider various aspects of the contributions of medium-sized and small firms toward innovation.

Part 3 considers the government components of the government–industry team in different countries including England, Japan, and the United States. These chapters are followed by a discussion of the role of government regulations and their effects on innovation; a prescription for experiments to improve government's role in management of technology; and, finally a remedy emphasizing the importance within government of a closer integration between scientific policy and economic policy.

Having considered an overview and each of the separate components, the final section blends the government and industry sections together. The first chapter in Part 4 suggests strategies for innovation to be used by government and industry. This chapter is followed by a description and plea for aligning industrial planning with government policy. The penultimate chapter reviews the evolution of scientific policy in the United States and urges a system that will inject industrial input into government science policy. It argues that the adversary roles between government and industry in relation to technological advancement is dysfunctional and can only impede progress. Cooperation and recognition of responsibilities between the public and private sectors will allow us to advance toward a world that is so easily within our grasp. The book ends on a short note that considers a subject mentioned earlier but only addressed obliquely, the relationship between technology and psychology. The note asks how we can select the directions of innovation and how people can trust in progress and not reject or withdraw from the opportunities it offers.

Many believe we can no longer allow science and technology to simply evolve. Similarly, we must be careful not to overdirect science and technology. It is important to balance and blend the individual responsibilities of industry and government in the future. If technological innovation is not focused, we can have the catastrophic consequences of mass, or at the very least, regional unemployment. There is, of course, the even more dire consequence of mass annihilation from the uncontrolled development of weapons of destruction or the consequences of innovations resulting in a spectrum of disasters from environmental health hazards to disposal of nuclear wastes. On the other hand, since innovations build on previous innovations and the science and technology is available, the opportunity exists to develop a world far more improved than the one we presently inhabit. Through careful blending of government and industry and thoughtful direction for technological innovation, regional unemployment differences can be decreased and the health and economic portions of our lives can be enriched. The choices are available; the paths remain to be chosen.

PART ONE

OVERVIEW:
ECONOMIC AND SOCIAL
ASPECTS OF INNOVATION

2 Present and Future Context for Innovation

K. L. R. Pavitt

Technological trends have a major influence on the future problems and possibilities of society, and they will contnue to pose policy problems and puzzles for industry and government, for economists and scientists. But predicting future technological developments and their interaction with policy is extremely difficult. In the short term, industrialists find it difficult to predict technological feasibility and the market's reaction to products embodying new technology. In the medium term, government officials are searching for better methods and mechanisms to support technologies that respond to society's needs. And in the long term, economic philosophers from Adam Smith and Thomas Malthus through Karl Marx and Joseph Schumpeter to John Kenneth Galbraith—all of whom saw clearly the major impact that technology has on economic change—have nontheless made assumptions about technology that have turned out to be wrong in significant respects.

Our basic understanding is such that we are still unable to predict with a high degree of confidence what the future impacts of technology will be.

In the past few months, I have considered the problems related to technology that could emerge in the Organization for Economic Cooperation and Development (OECD) area in the next 20 years. The approach that I adopted inevitably had many limitations. I did not develop any rigorous, explicit, and coherent model of the interaction of technology with various aspects of policy; given our lack of fun-

This paper is based on a longer one prepared by the author, on the future of technology in advanced industrial societies, for the Interfutures programme of the Organization for Economic Cooperation and Development, Paris. The views expressed in this paper are those of the author.

damental understanding, I preferred to draw eclectically on insights from a variety of approaches. Nor did I discuss some important subjects: the specific attributes of military technology, technology for the less-developed countries, and purely social aspects of technology such as computers and secrecy, bugging, and mind-influencing drugs. It can be argued—perhaps correctly—that it is in these areas that technology will have its major impacts in the future. However, I doubt that these areas can be divorced from the underlying relationship among industry, technology, and economc performance, on which I concentrate here. I emphasize potential problems rather than potential achievements, not to encourage a fashionable doom and gloom, but in the belief that problems are more likely to be solved if they are detected, discussed, and analyzed well in advance. And I make assertions not on the basis of any cool and careful calculation of probabilities that I do not believe anyone can do, but in the hope of provoking discussion and debate.

Essentially, I am trying to answer three questions. Will there be any technology-related hindrances to overall economic growth within the OECD area? What international problems are likely to exist because of technology-related activities? What shifts can be expected in the sectoral pattern of technical change?

Technical Change and Economic Growth

The diffusion of technology takes a long time. Within each manufacturing sector in the industrialized countries, best-practice firms embodying the most recent technology often have productivity levels two or three times higher than firms using older technologies. Thus what is already in the research and development pipeline means that manufacturing productivity levels at least three or four times higher than those existing today will be feasible in the near future. This means that the technology supply per se should not be an obstacle to the continuing growth of manufacturing in the next 20 years. Nonetheless, a related series of problems could emerge.

The first concerns the provision of inputs to expanding the manufacturing industry, namely materials, energy, and environmental goods. The provision of energy could pose problems unless there is vigorous technical change in methods of energy conservation and in developing new energy sources. For better or for worse, governments and especially the U.S. government, will have a critical influence on this process, especially through their research and development funding and their pricing policies.

The second series of problems relates to the changing economic climate since the beginning of the 1970s. Technical change thrives in a buoyant economic climate. High rates of investment lead to high rates of use of new techniques; optimistic expectations about the future generally result in high levels of research and development and other innovative activities. Stagflation, the combined phenomenon of stagnation and inflation, will certainly slow down the rate of diffusion of new techniques. It might eventually have an effect on the level of research and development and other innovative activity financed by industry. So far, it has not resulted in any noticeable decline of company-financed research and development in OECD countries, although information for the United States suggests a noticeable shift toward short-term product improvement at the expense of long-term radical change.

This means that the widely accepted function of government in financing longer-term and higher-risk research activities will be even more important in the future than it was in the past. And herein lies a third possible problem area, namely future government policies for the funding of research and developing activities. Support for small-scale basic research activities has become unfashionable since 1965, in part because of the intrinsic difficulties of measuring their economic and social utility. In times of budgetary restraint, such programs may be squeezed out more easily than those organized around big machines (e.g., satellites, accelerators, telescopes), the social value of which could well be much smaller.

Similarly, some governments continue to support programs in the so-called high technologies (aerospace, nuclear energy, advanced electronics) on a very large scale. Although there is some justification for programs in these sectors, decisions about them are generally taken in secret, and the criteria for government support have been dominated by nationalism and a desire for technological autarchy, rather than by a realistic assessment of costs and economic and social benefits. The result has often been large-scale government support for projects that were commercial failures and the distortion of national research and development priorities. In a future world of monopoly power based on resources and technology of discontinuities and of the threat and the reality of unemployment, the pressure for such programs is unlikely to decrease.

Technical Change and Economic Performance Relations Among Industrialized Countries

Since 1960, there have been significant and major shifts in economic performance and power among the OECD member countries. Most of these changes are closely correlated with performance in the manufacturing industry, as measured by trends in labor productivity and world export shares. This performance in turn has been closely related to the ability to embody new technology in production systems and in products, although it is difficult to separate the technology from other nonprice factors influencing economic performance. Nonetheless, trends in technological activities in the industry of different countries may give some indication of future trends in industrial performance.

These trends show that the extent of the technological gap between the United States and the rest of the OECD, which was of major concern during the 1960s, has been exaggerated. Aside from the United Kingdom many other major OECD countries (the Federal Republic of Germany, Japan, the Netherlands, and Switzerland) increased their industry-financed research and development as a percentage of GNP more than the United States. This naturally leads to speculation about a possible future technological decline in the United States. Signs of this are already visible in certain capital goods, and a number of writers have given reasons why it could happen: the draining of the technological resources of the United States from the capital goods sector by large-scale military and space programs; the availability of easy profits from overseas investment and sales in client states instead of industrial and technological regeneration at home; and a return to a long-standing German predominance in industrial technology. Available evidence does not show convincingly that a decline is happening in the United States, only that it could. If it did, the national and international consequences would be considerable. For example, any loss of jobs and of international influence could be counteracted in the short term by increases in defense research and development and procurement, but this would only aggravate the long-term process of decline.

Any discussion of future industrial decline in the United States is speculative and controversial. The problems posed by divergences in industrial and economic performance in Western Europe between France, Italy, and the United Kingdom, on the one hand, and the Federal Republic of Germany, the Netherlands, Belgium, and Switzerland on the other are far more immediate and real. It is widely recognized—among others, by the President of the EEC Commission—that continuing divergence in the future will lead to increasing tensions and to the eventual break-up of the Common Market. The conventional economic prescriptions put forward to resolve this problem are monetary discipline, wage-claim discipline, and general calls for greater efficiency. Although necessary, such measures are unlikely to be sufficient, given that the Federal Republic of Germany and some neighboring countries continue to be able to sell their capital goods in international markets in spite of a very considerable price disadvantage. This in part reflects the higher technical quality of capital goods made in the Federal Republic of Germany, Switzerland, and the Netherlands and is no doubt related to the traditions of thorough training and high status given to engineers in these countries. Economic convergence and stability in Western Europe will probably require that France, Italy, and the United Kingdom absorb something of these traditions.

An overall relative decline of United States industrial technology could cause one set of problems, and the strong lead of the United States in the high technologies (aerospace, nuclear, advanced electronics) could cause others. American firms and the United States government have strong monopoly power in the provision of aircraft, rocket launchers, nuclear fuels, computers, and advanced electronic components. The United States government is naturally conscious of the international diffusion of some parts of these technologies. The OECD countries are conscious of the critical importance of certain of these sectors for civilian industry and are naturally reluctant to remain dependent on a foreign monopoly supplier, however well intentioned.

In relation to space technology, the problem has to some extent been resolved as a result of the dilution of the United States' monopoly, because of the participation by other countries in Intelstat, which is the controlling policy body for the western world's international satellite system. However, space technology is not considered as economically important as it was 10 to 15 years ago, which is another reason why less is heard today about any "problem" in this area. This is certainly not the case in relation to nuclear fuels. For the last 20 years, the United States govrnment has effectively prevented any substantial international transfer of its technologies for enriching and reprocessing nuclear fuels, with the result that countries such as France, the Federal Republic of Germany, and Japan have all obtained the technologies through independent programs of redevelopment. Today, the concern of President Carter's administration with nuclear proliferation has led it to try to influence nuclear fuels policies of other countries in directions that these countries consider incompatible with their long-term energy requirements.

Thus the experience of the past 20 years suggests that these problems will not go away quietly in future. Governments in many countries will continue to support large—and sometimes wasteful—programs in the high technologies, and these technologies will continue to be the subject of international negotiation and debate. There will probably be increasing talk about "internationalization" as a solution to the problems of dependence, and many of the proposed prescriptions will (as in the past) be ambiguous. International cooperation between the United States and other countries will be satisfactory to the latter insofar as it gives them effective control over the use of the technologies, and to the former insofar as it does not. International cooperation among countries other than the United States to reduce the burden of high-technology programs will be efficient insofar as there is an effective division of technological labor within them, but this will mean the replacement of one form of technological dependence by another.

Problems of technological dependence could also emerge in relation to the Soviet Union. In spite of the allocation of considerable resources to education, investment, and research and development activities, the U.S.S.R. continues to lag behind the OECD area in the introduction and diffusion of industrial technology. In spite of considerable efforts to overcome these difficulties (for example, by bringing research and development activities closer to production, by introducing economic incentives, by importing Western technology), they are likely to persist as long as prices of capital goods are determined bureaucratically and research and design teams are organized hierarchically.

The persistence of these problems, coupled with the continuing Soviet importation of Western technology, could lead to important policy changes in the future. On the Soviet side, any continuing poor industrial and technological performance could lead to a reappraisal of the policy of importing Western technology. On the Western side, continuing Soviet technological dependence could lead to the exertion of stronger political pressures. But such changes in policy would be based on an exaggeration of the importance of West–East technology transfer. In and of itself, it cannot solve the structural problems of the U.S.S.R., and the U.S.S.R. could survive without it, if necessary.

Technical Change in Different Industrial Sectors

It is very difficult to make any simple generalizations or speculations about the direction of future technical change. The factors influencing it are many and diverse, and their influence will vary among industrial sectors. In addition to scientific and technological opportunity, they include movements in wage rates, industrialization of today's underdeveloped countries, the cost of resource inputs, changing consumer attitudes and demands, and changing worker attitudes. Broadly speaking, they will pose problems for some groups of industries and opportunities for others.

Process industries, electrical generation, and transport equipment (in this case aircraft and ships) have certain common characteristics. They all offer considerable opportunities for exploiting economies of scale, and this has dictated the main directions of technical change over the past 20 years. However, for a mixture of environmental, technical, and economic reasons, the potential for exploiting economies of scale in the future will be more limited.

These industries are all big consumers of energy and increasing energy costs means that both the rate and the direction of technical change in the future will be very different from in the past. In terms of labor productivity, the rate of future change is likely to be lower. However, given higher energy prices and potential oil shortages, there will be considerable technical opportunities and challenges open to the chemical industry (coal gasification and liquefaction) and to the electrical industry (nuclear, solar, and wind).

A number of factors are likely to influence the future pattern of technical change in durable and other consumer goods: market saturation in the industrialized countries; increasing consumer concern with quality, safety, and reliability, which is likely to reduce the volume of the replacement market and to restrict the opportunities for new-product development; and increasing competition from production in the less-developed countries.

Clues about possible new areas of product development can be detected in the trends in consumer-related expenditures. With increasing personal income, a growing proportion is spent on education, health, and housing. If one accepts (as this writer does) that, with increasing wealth, there is a movement toward a self-service economy, where essential demands are met increasingly in a capital-intensive—rather than a labor-intensive—fashion, then the possibilities of technical change appear to be considerable. Several examples include the consumer-operated video systems for education; a noncancerous replacement for the cigarette; self-diagnosis equipment; and increasingly capital-intensive construction models. However, given the nature of these technological innovations, it is not certain whether they will originate or be controlled by the traditional suppliers of consumer goods. They will certainly require major social and economic changes including overcoming resistance from powerfully entrenched interests, if they are to be introduced on a large scale.

The manufacturing costs of traditional consumer products will also create problems. On the one hand, there will be increasing pressures of competition from the less-developed countries; on the other hand, there will be increasing workers' aspirations for a more satisfactory working environment.

Innovations in machinery and other production equipment will continue to be significant. The supply of technology applicable to machine building has increased considerably in the period since 1945. In addition to the mechanical sciences, industry has learned to incorporate and use advances in materials science, hydraulics, hydro- and aerodynamics, and electronics, so that the productivity, reliability, and quality of machinery has increased considerably. Textile machinery, for example, has become much more sophisticated, under pressure of competition from production with standard machines in the less-developed countries. At the same time, a company such as Volvo has responded by abolishing the traditional assembly line, by creating small work groups, by using job rotation and self-inspection, and by trying to increase the involvement of the entire work force. The Volvo experience is updated and discussed in a bit more detail later in this volume.

However, it remains to be seen whether both challenges can be met at once. If they cannot, then the result is likely to be the fully automatic factory or production in the less-developed countries. In either case, the implications for the advanced countries will be the same. An upgrading of skill and employment requirements in the capital-goods sector, coupled with huge employment losses and a less-skilled working force required in the sector making consumer goods.

Conclusions

What conclusions can be drawn from this brief review of potential future policy problems regarding technological change?

First, problems that have been with us will remain with us for a long time: in particular, the problems of designing effective systems of government support for critically important civilian technologies and of working out equitable and safe international controls in areas such as nuclear fuels and rocket launchers. Second, some government policies will be more difficult to implement than in the past, in particular, providing a macroeconomic climate favorable to technical change while concomitantly supporting long-term basic research. Third, some new problems may turn out not to be problems at all; for example, the fear of a reduction of research and development and the innovative activities of industrial firms.

Finally, some very old spectres remain to haunt us: the Malthusian fear of diminishing returns in energy investment and the Marxian fear of technology-induced unemployment on a large scale. It is no comfort that both predictions were both wrong in significant respects in the past. Structural and technical change will be as rapid as it ever was, but very different. Economic and social progress will depend on our ability to understand and manage these changes, whether at the level of the firm, the nation, or the international community. The rest of this volume addresses these issues with the objective of increasing the focus and presenting alternatives for concerned citizens and decision makers in government, industry, and academia.

3 Returns from Industrial Innovation

Edwin Mansfield

This analysis summarizes very briefly the results of several studies carried out by my colleagues and me during the past few years. These studies are concerned with measuring the social and private rates of return from investments in new industrial technology. Economists have long stressed the need for more detailed and direct measures of such rates of return, since they obviously play a major role in the rational formulation of public policy toward civilian technology. Yet, with the exception of a few agricultural innovations, no such measurements have been attempted. To help fill this gap, we have carried out a detailed study of the returns from 17 industrial innovations.

The innovations we studied occurred in a variety of industries, including primary metals, machine tools, industrial controls, construction, drilling, paper, thread, heating equipment, electronics, chemicals, and household cleaners. They occurred in firms of quite different sizes. Most of them are of average or routine importance, not major breakthroughs. Although the sample cannot be regarded as randomly chosen, there is no obvious indication that it is biased toward very profitable innovations (socially or privately) or relatively unprofitable ones.

Moreover, we obtained very rich and detailed data concerning the returns from the innovative activities of one of the nation's largest firms from 1960 to 1972. For each year, this firm has made a craeful inventory of the technological innovations arising from its research and development and related activities, and it has made detailed estimates of the effect of each of these innovations on its profit stream. On the basis of these and other data, we computed the private rates of return and lower bounds on the social rate of return, from this firm's investment in new technology. Among other things, these results provide a valuable check on the results of the 17-innovation sample.

To estimate the social benefits from an innovation, we used a model that was similar to the one employed in earlier agricultural studies, but it was necessary to modify and extend it in several ways. This model is described in detail in a recent paper in the *Quarterly Journal of Economics*, as well as in our book.[1] Our major empirical results can be summarized as follows: First, the social rates of return from the investments in the 17 innovations tend to be very high. Specifically, the median estimated social rate of return is about 56%. (And, for a variety of reasons, these estimates are likely to be conservative lower bounds.) Further, the estimated lower bound on the social rate of return from the major firm's investment in new process technology between 1960 and 1972 was also about 50%. To put these results in perspective, recall that Griliches found that the social rate of return from hybrid corn—a very successful innovation —was 37%.[2] Clearly, the investments that have been made in industrial innovation, on the average, have yielded handsome social returns, if these innovations are at all typical.

Second, the private rates of return from the investments in these innovations seem to have been much lower than the social rates of return. The median private rate of return, before taxes, was about 25%. In interpreting this number, it is important to recognize the riskiness of this kind of investment and the fact that these are pretax figures. This riskiness is evidenced by the enormous variation among the 17 innovations in the private rate of return. In the case of six innovations, the private rate of return before taxes was less than 10%, whereas for five innovations, it was more than 40%. Also, the private rate of return from the major firm's total investment in innovative activities from 1960 to 1972 was 19% before taxes, which is not too different from the median private rate of return already discussed.

Third, in about 30% of the cases, the private rate of return was so low that no firm, with the advantage of hindsight, would have invested in the innovation, but the social rate of return from the innovation was so high that, from society's point of view, the investment was worthwhile. When we look at specific innovations, the difference between the social and private rate of return seems to be directly related to the economic importance of the innovation, measured by absolute annual benefits, and inversely related to the cost of imitating the innovation. These results are quite consistent with hypotheses put forth by Arrow et al.[3]

18 Overview: Economic and Social Aspects of Innovation

Fourth, for nine of these innovations, we could obtain data concerning the approximate private rate of return expected from the innovation by the innovator when he began the project. In five of the nine cases, this expected private rate of return was less than 15% before taxes, which indicates that these five projects were quite marginal from the point of view of the firm. Indeed, the executives of the firms confirmed that they were considered marginal. Yet the average social rate of return drops precipitously when the expected private rate of return falls from 10 or 12% to, say, 5 or 6%, this result seems to suggest that there may be many projects where the expected private rate of return was a bit lower than for these five projects, with the result that they were not carried out, but where the social rate of return would have been quite high.

Fifth, detailed studies of this sample of innovations show the key role of competiton in driving a wedge between private and social returns from innovating. In some of these cases, competitors began producing (or using, if the innovation was a process) the innovation within a few years after the innovation was first commercialized; and because the innovation could not be patented or the patents were weak, the innovator could not prevent imitation. Confronted with such competition, there was no way that the innovator could appropriate the lion's share of the social returns from the innovation.

Finally, what are the policy implications of these studies? The work summarized very briefly here was carried out to shed light on some relatively fundamental topics in the economics of technological change that are important both to the development of more adequate models and measurements and to the formulation of more effective public policy. Since these studies were not designed to yield results with direct relevance to specific policy questions, it would be unrealistic to expect them to lead to very definite or concrete policy proposals. Moreover, in view of the limitations of these studies, we must be careful to avoid putting more weight on them than they realistically can support. Certainly, all that these results can provide are tentative indications, or clues, concerning certain general policy questions in this area. However, since so little is known concerning these questions, it may be worthwhile to present a few of these indications or clues, tentative though they must be.

Although our sample is too small to support definitive conclusions, the results certainly suggest that, even taking into account the riskiness of innovative activity, the social rate of return from investments in new technology has tended to be very high. If the marginal social rate of return from investments in civilian technology is greater than the marginal social rate of return from other uses of the relevant resources, this is evidence of an underinvestment in civilian technology. Fellner and Griliches have argued that it is legitimate, at least not too rash, to make the jump from average to marginal rates of return for investments of this sort.[4] Indeed, in Griliches' view, there is no reason to believe that the marginal rate of return differs much from the average rate of return. If this is the case, our results certainly suggest that there may be an underinvestment in civilian technology in the United States, since the average rate of return seems very high.

At present it is difficult, if not impossible, to specify what combination of measures, such as increased federal grants or contracts and research and development tax credits, would be most effective in compensating for whatever underinvestment there is in civilian technology. However, it seems reasonable to believe that, if a program along this line is deemed worthwhile, it should be neither large scale nor organized on a crash basis; it should not focus on helping sick industries merely because they are sick; it should not get the government involved in the larger stages of development work; a proper coupling should be maintained between technology and the market; and the advantages of pluralism and decentralized decision making should be recognized. Given the current uncertainties, it would seem wise to proceed with caution and to try to resolve many of the key uncertainties before too big a commitment is made. After all, the federal government, in trying to improve matters, could do more harm than good.[5]

In conclusion, much more work is needed in this area both to extend and replicate existing studies and to carry out new ones. The work carried out to date does little more than scratch the surface. Fortunately, government agencies recognize this fact and are making a number of attempts to extend the relevant data base and improve the methodologies used. In particular, the National Science Foundation is currently supporting a number of projects in this area. Also, the National Aeronautics and Space Administration and the Departments of Commerce and Labor have been carrying out work of this sort. In addition, several groups, such as the Office of Technology Assessment of the United States Congress and the Organization for Economic Cooperation and Development, have organized committees to study these and related topics. There is every reason to believe that knowledge in this area will expand and deepen in the next few years.

4 The Effects of Technological Change on Employment

Roy Rothwell

The large majority of workers are employed in the basic, traditional industries even in the so-called "higher technology" countries. It is with this fact in mind that an analysis is presented on the effects of technological change on employment in the textile industry.

Since 1945 textile machinery has undergone a technological revolution. Not only has the pace of technological change increased greatly, but much of the change has become technically more radical in nature, often embodying techniques from other areas such as electronics, aerodynamics, and chemical technology. This technological revolution has had profound implications for employment in both the producer and user industries. It has forced textile machinery producers to acquire personnel with higher levels of technical expertise than previously required. In addition, it has greatly reduced the manpower component of textile production for textile machinery users. This chapter briefly presents empirical data that illustrates the effect of technical change on the quality of manpower required in the producer industry and on the levels of manpower employed in the user industry.

Innovation and Manpower

The Textile Machinery Industry. A recent detailed study of innovation in textile machinery has clearly shown that, in general, those companies that have been "technically progressive" have survived and prospered during the post-Second World War years.[1] In contrast, other companies have suffered drastic decline during this period from positions of eminence to positions of near oblivion today primarily because they failed to update sufficiently existing models and to produce new generations of machinery. Very strong evidence also suggests that the technical sophistication of textile machinery plays a dominant role in determining its export competitiveness.

This evidence, although illustrating the importance to commercial success of technical change per se, fails to indicate the relative importance of the different types of technical change—large-step, "radical" change, and small-step, "incremental" change. A number of instances can be quoted where the introduction of a radical innovation has led to the rapid growth of new firms or new divisions within a firm. Probably the most famous examples of this are the establishment of the enormously sucessful Sulzer Weaving Section of Sulzer Brothers after the revolutionary Sulzer Flying Gripper Weaving Machine was introduced, and the rapid growth of the Kovoslav National Corporation in Czechoslovakia after its commercial introduction of the new open-end spinning technique. In most areas, in fact, short-term prosperity can be insured through simple "product improvement" innovation. In the longer term, more radical innovation is necessary to maintain a firm's competitive position and foster growth.*[2] To summarize, then:

1. Technical change is an important determinant of the commercial success of a firm and of the export competitiveness of textile machinery.
2. Radical technical change has become increasingly important in ensuring a firm's commercial success and long-term survival.

But what are the manpower implications of this? Are there significant manpower-related differences in the factors associated with the production of successful incremental innovations? Yes, there are.

Difference in Technical Manpower Between Radical and Incremental Innovators. Table 4-1 shows the averaged characteristics of the development departments in 20 companies that were success-

---

*This does not mean that radical and incremental change are mutually exclusive. On the contrary, they are often interrelated and complementary, and a radical change will spark off an extended series of incremental changes. The point is, however, that without the initial major step, the series of minor steps could not have occurred.

Table 4-1. Characteristics of the Development Department—Radical and Incremental Innovators

Question	Response	Classification	
		Radical	Incremental
Does the firm have a formal research-and-development department?	Yes	14	2
	No	6	13
If not, does the firm have a formal design-and-development department?	Yes	5	1
	No	1	2
Does the firm have:			
a. Both		6	1
b. Neither		1	2
		Concentration (%)	Concentration (%)
How many nongraduate engineers were associated with development work at the time of the project? (average no./firm)		6	3
		0.8	1.3
How many graduate engineers were associated with development work at the time of the project? (average no./firm possessing graduates)		3.4[a]	1.7[a]
		0.45	0.75
How many graduate scientists were associated with development work at the time of the project? (average no./firm possessing graduate scientists)		2.2[b]	2.0[b]
		0.3	0.88

[a] These are the average numbers for firms that actually employed graduate engineers. Two radical innovators and seven incremental innovators did not employ a single graduate engineer in development.

[b] These are the average numbers for firms that actually employed graduate scientists. Seven radical and 13 incremental innovators did not employ a single graduate scientist in development.

ful in producing radical textile machinery innovations (radical innovators) and in 15 companies that produced only small-step innovations (incremental innovators). The radical innovators had an average of three times the employment of incremental innovations.

The data presented in Table 4-1 suggest that successful radical innovations are associated with a formal research and development department staffed by graduate-level engineers or scientists, particularly the former. Incremental innovations, in contrast, tend to occur in a design-and-development department staffed primarily by nongraduate engineers.* The research and development will, of course, be capable of producing incremental as well as radical innovations. Without expert external assistance, however, the design and development department will generally be unable to cope with radical technical change. Moreover an analysis of eight unsuccessful radical innovations showed that in seven cases, the lack of in-house technical expertise resulting in the inability to solve technical problems made a major contribution to failure. Further, several companies quoted a lack of skilled technical manpower as posing the major barrier to their endeavors to innovate.

Policy Implication. Since the theme of this meeting is "Government/Industry Cooperation for Technological Innovation," it is relevant at this point to attempt to assess the policy implications of this data. The major policy implication is clear. That is, since radical innovation is important to the commercial success of textile machinery companies, some way must be found to inject the necessary high level of technical expertise into those companies—particularly smaller companies—that currently do not possess this expertise.*

Simply stepping up the supply of graduate engineers would probably have little impact on the situation, particularly in the United Kingdom where a number of graduate engineers have recently been

---

* In both kinds of companies the chief executives often had technical qualifications—70% for radical innovators, 60% for incremental innovators. Of the radical innovators 55% were qualified up to chartered engineer level, including 35% graduates, whereas in the case of incremental innovators only 33% were qualified to this level, including 26% graduates.

* The proposition that differences in technical manpower levels explain the superior performance of the West German textile machinery industry over its counterpart in the United Kingdom is an attractive one, as the following figures suggest: In West Germany the percentage of total employees who are "technical employees" in 1974 equaled 16.4%. Ten percent of employees are occupied in "R and D and Construction." In the United Kingdom the concentration of scientific, technical, and drafting personnel in 1973 equaled 5.6%.

reported as unemployed anyway, and where a number of textile machinery companies have complained of difficulties in attracting graduates to work for them. The textile machinery industry in the United Kingdom, it seems, has an image that is unattractive to graduates. Some smaller companies also feel that they cannot afford to employ graduates. In view of these results, however, it might be that they cannot afford not to employ such people! A more practicable proposition would probably be for the government to assist with training schemes designed to generally upgrade the technical knowledge of technicians currenlty employed in the industry and to support specialist courses in textile technology for them.

Perhaps governments can best assist through policies relating to technology transfer from specialist units. This policy has paid handsome dividends in Czechoslovakia where there is a total government effort in several specialized institutes of over 1100 technically qualified personnel, a high percentage of whom are graduates. This effort enabled the Czechs to market their BD 200 open-end spinner 4 years ahead of the major Western competitors. They now have about 70% of the total world market in open-end spinners, and they seem set to achieve the world lead in multiphase weaving.

Finally, technical innovation, and particularly radical innovation, is a risky business, and even firms containing a comprehensive range of technical skills might not possess a sufficiently sound financial base to enable them to take the high risks involved in producing a revolutionary new machine. Perhaps governments could assist firms with selected developments that are likely to provide the nation with a technical lead, or that could cut down significantly on imports, via the provision of risk capital.

The Textile Industry. The principle development trends in textile machinery during the post-Second World War years have been geared toward three primary ends: 1) increasing the productivity of individual units; 2) reducing the number of operations required in a particular process; and 3) increasing the amount of automatic transfer between operations. Points 2 and 3, in particular, have had the objective of reducing manpower requirements in the mill. This has been especially important in the developed countries where, for many years, there existed a shortage of skilled operatives, and where relatively high wage levels rendered labor-intensive processes uncompetitive in the face of increasing texitle production from expanding textile industries in the low labor-cost developing countries. As a consequence, the textile industries in the advanced economies have changed from labor-intensive craft industries to the capital-intensive, machine-oriented industries of today. This is demonstrated for the weaving process in the United Kingdom in Table 4-2.

Increases in labor productivity in the spinning and weaving industries in the United Kingdom between 1950 and 1970 are shown in Table 4-3. A high proportion of these increases in labor productivity

Table 4-2. Percentage of Value Added in Weaving

	Labor cost (%)	Capital cost (%)	Power, building, profit, etc. (%)
1955	68	20	20
1970	25	70	5

Table 4-3. Output, Employment, and Productivity in the Spinning and Weaving Industries in the United Kingdom (cotton and man-made fibers)

	1950	1960	1970
Spinning			
Output (million lb)	943	640	441
Employment (thousands)	107	63.7	32.6
Productivity (million lb/thousands)	8.8	10.04	13.53
Weaving			
Output (million linear yd)	2830	1911	1216
Employment (thousands)	129.2	81.3	43.0
Productivity (million yd)	21.9	23.5	28.3

can be ascribed to the increased use of new and improved textile-processing machinery.

Changes in the relationship between human effort and textile production as a result of technical change are graphically illustrated in Figure 4–1, which shows the change in human effort involved in reproducing the same quantity of yarn and of cloth from 1750 to 1950. Clearly, technological change has had a significant impact on the labor content of both.

The remainder of this paper concerns itself with describing the

26 Overview: Economic and Social Aspects of Innovation

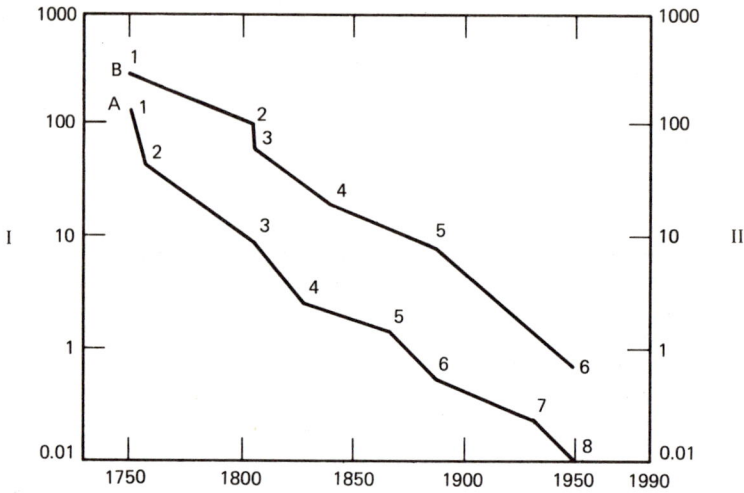

Figure 4-1. The change of human effort in the course of time.

detailed relationship between technological change and changing manpower requirements in the textile industry. Since it would be too lengthy a procedure to cover all aspects of textile production, the paper will concentrate solely on the staple yarn spinning sequence. A schematic of the cotton spinning process is shown in Figure 4-2 which briefly describes the function of the various machines involved.

The next stage of the prespinning process (see Figure 4-2) was at a much less satisfactory stage of development. It involved:

1. Carding the lap, 1 yd of which weighed about 1 lb, to produce a sliver, of which about 100 yds weighed 1 lb. The production rate of one carding engine (card) was about 8 lb/hr and the sliver was delivered coiled into cylindrical canisters each holding about 10 lb of sliver.

2. Drawing the sliver through a succession of accelerating rollers in order to parallelize the fibers before converting them into roving. Typically six slivers from six cans are fed into each drawing unit. The draw ratio is about six so that the delivered sliver has roughly the same weight per unit length as the individual input slivers and differs only in the degree of orientation of its component fibers. The production rate of each

drawing unit is about 100 lb of sliver/hr, delivered in cylindrical cans exactly like those used at the card. To achieve the requisite degree of fiber orientation, all sliver usually passed through three successive stages of drawing.

3. Production of roving from sliver. This was done on spindle and flyer frames fed from 10-lb cans of sliver to produce roving on bobbins weighing from 1 to 3 lb depending on the fineness of the yarn to be produced by the spinning frames. A typical roving frame would have from 100 to 150 spindles.

A visitor to an early 1950s mill would have been immediately struck by the atmosphere of the early industrial revolution of this stage of yarn manufacture. In the smallest mill equipped with only

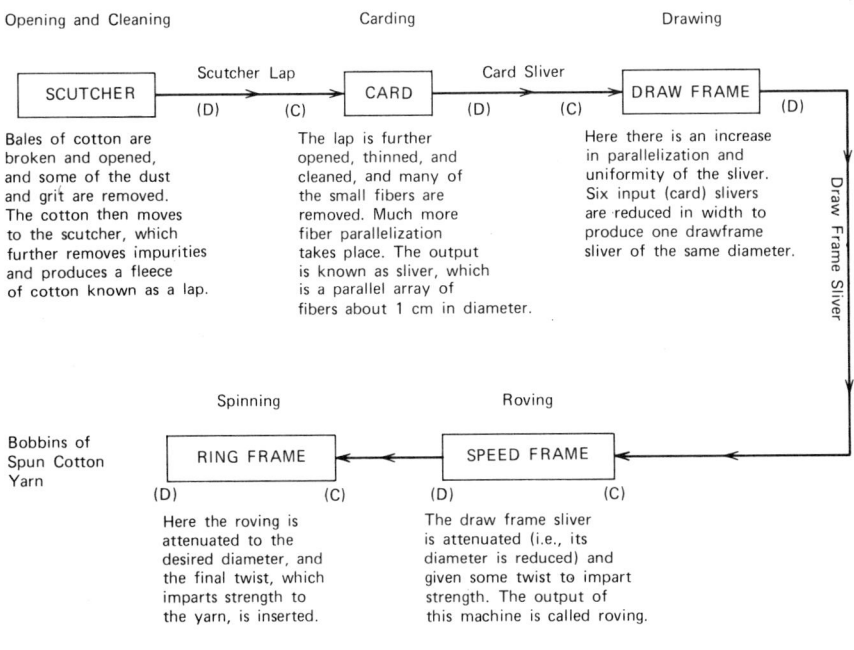

D — Take—off or doffing point.
C — Feed—in or creeling point.

Figure 4–2. The cotton spinning process.

28   Overview: Economic and Social Aspects of Innovation

one opening line (1000 lb/hr) a visitor would have seen 125 cards—each a large machine occupying about 100 ft^2 of floor space—with a partially exhausted 40-lb lap of cotton on its feed table and a partially filled 10-lb sliver can under the delivery coiler. Downstream from the cards would be three sequences of drawing each consisting of 12 to 14 drawing units, each fed from six sliver cans and delivering into a seventh. Downstream again from the draw frame would be the spindle and flyer frames with a total of 250 to 500 spindles each fed from one of the sliver cans delivered by the final draw frames. The input and output of these processes are shown in Table 4-4, along with the corresponding operative hours per unit of production (OHP) figures.

Perhaps the dominant features of this mill scene were the amount of feeding material into (creeling) and removing material from (doffing) the machinery that was being done manually, and the number of cans of sliver about the place. In addition to buffer stocks of full and empty cans, there were 725 cans actually in use each, at any one time, holding on average 5-lb sliver and each representing a creeling or doffing point requiring more or less frequent manual attendance. At the cards it was necessary to creel in 25 laps/hr and doff 100 cans/hr. The drawframe section presented the scene of most frenzied activity, requiring creeling in 300 cans/hr and doffing a similar number, each creel can remaining in position for about 42 min and each of the 36 delivery stations requiring the doffing of a can approximately every 7 min. The spindle and flyer-frame section was relatively quiet with 350 cans that remained in place for about 3.5 hr.

Table 4-4. Input and Output of Successive Processes and OHP

Operation and Machinery Requirement	In	Out	OHP
Carding—125 carding engines	125×40-lb laps	125×10-lb sliver cans	0.703
First Drawing—12 drawing units	72×10-lb sliver cans	12×10-lb sliver cans	
Second Drawing—12 drawing units	72×10-lb sliver cans	12×10-lb sliver cans	0.547
Third Drawing—12 drawing units	72×10-lb sliver cans	12×10-lb sliver cans	
Roving—350 spindle and flyer units	350×10-lb sliver cans	350× 2-lb sliver cans	1.288
Cardroom General	—	—	0.273
			3.16

The Effects of Technological Change on Employment

It is difficult to give a clear yet concise picture of the tasks involved in marshaling, doffing, and creeling sliver cans. The cans themselves were generally 9 or 10 in. in diameter and about 30 in. high—not by any means ideal for handling but necessarily of that shape for proper functioning of the coiling mechanism and the stacking requirements for a satisfactory creel. Doffing required only removing the full can and then deftly replacing it with an empty can and severing the sliver. Creeling required timely removal of the emptying can, replacement by a full can, and neat and skillful splicing of the new sliver to the tail of the old one.

Obviously this situation had much scope for the introduction of technological change to reduce the amount of manual handling of material as well as to increase the productivity of individual production units.

Technological Changes in Spinning, 1950–1968. The main technological changes that have taken place that had the most marked impact on the prespinning sequence described previously are, briefly:

1. The increase in card production rates by a factor of about 10, with direct feed from card to drawframe.
2. The increase in drawing speeds of the same order as the increase in carding rate. With the old card and drawframe speeds, a 10-lb package was near the optimum. After the 10-fold increase in speed, cans of 80–120 lb could be used.
3. The improvements in drafting performance and the precision of the drafting process reduced sliver variability to such an extent that it enabled the number of drawing sequences to be reduced from three to two.

The machinery line-up in a modern (1968) mill is shown in Table 4–5.

It is immediately obvious from Table 4–5 that the amount of package handling has been drastically reduced. No creeling is now needed at the cards, and the doffing requirement has been reduced from 100 cans/hr to 10. The drawframe creeling requirement has been reduced from 300 cans/hr to 20, and drawframe doffing reduced from 300 cans/hr to 25. This, coupled with the increases in production rate of the individual units, reduced the OHP from a total of 3.16 in 1950 to 1.014 in 1968.

The impact of this technological change is probably best summarized in Tables 4–6 and 4–7, which give complete details of ring spinning mills for 20S and 30S cotton in 1950 and 1968 respectively.

Table 4-5. Modern Machinery Line-up and OHP

Operation and Machinery Requirement	In	Out	OHP
Carding—14 carding engines	Direct feed	14×100-lb sliver cans	0.078
First Drawing—4 drawing units	32×100-lb sliver cans	4×100-lb sliver cans	0.156
Second Drawing—4 drawing units	32×100-lb sliver cans	4× 40-lb sliver cans[a]	
Roving—250 spindle and flyer units	250× 40-lb cans	250× 31-lb sliver cans	0.468
Cardroom General	—	—	0.312
			1.014

[a] The use of smaller cans for sliver for the delivery from the second stage of drawing is dictated by economic considerations in the design of the flyframe creel.

The total number of direct operatives has been reduced from 78 in 1950 to 29 in 1968 for 20S cotton, and from 99 in 1950 to 35 in 1968 for 30S cotton. The corresponding figures for indirect labor are 22 and 16.5, respectively (20S cotton), and 25 and 19.5, respectively (30S cotton).

Clearly between 1950 and 1968 technological change had a major impact on labor requirements in the conventional cotton spinning mill. It is interesting to note that the nature of the technological changes that contributed to this siuation were exclusively incremental ('small-step" innovations) in nature.

Post 1968. Although contributions to increases in the productivity of the prespinning sequence and to a reduction in the manpower component of this aspect of textile production were brought about through incremental technical change, the greatest change in the spinning process itself was the result of a radical innovation, the introduction of open-end spinning.

Table 4–8 compares the economics of short-staple yarn production for a 1945 spinning mill, a 1963 spinning mill, and a modern open-end spinning mill. It demonstrates the move toward greater machine use—from one shift working in 1945 to four shirts working in 1973—which accompanied the introduction of more modern machinery and which made a significant contribution toward the better economics of yarn production. The table also shows that unit labor costs were reduced by one-half between 1945 and 1963, and by one-half again between 1963 and 1973. These improvements could not have been achieved simply through the use of automated processes,* or by improved materials handling since, even though these would have reduced the work force, they would also have increased machinery costs, whereas the number of machines would have remained constant. The real advances were made in the production rate of machines and, as limits were being approached in ring spinning, open-end spinning was developed that offered a threefold improvement in productivity per rotor.

Radical technical change has also had a marked impact on the manpower requirement in long-staple (worsted) spinning. Comparison between a best-practice modern conventional worsted spinning mill and a mill using the radical new REPCO self-twist system is made below in Table 4–9. This shows that, using the REPCO system, OHP are reduced by almost a third from 3.68 with the conventional sequence to 2.67 with REPCO.

Other Factors

Rationalization and Concentration. Technical change, of course, is not the only factor that has had an impact on labor productivity and manning levels in the textile industry. Greatly increased concentration since 1945, with the formation through rationalization of large, integrated, capital-intensive production units, has led to the more efficient use of labor with a consequent reduction in employment.

An interesting example of this is the case of Carrington Viyellas' reorganization of Combined English Mills, which illustrates the complex interaction between the use of up-to-date machinery, product mix, market diversity, management style, and employment.[4] Combined English Mills (CEM) had taken full advantage of the re-equipment scheme of the government of the United Kingdom. By 1964, when the mill was acquired by Viyella, few mills had re-equipped to a greater extent than CEM's mill at Leigh. CEM did, however, suffer from the following faults: excess capacity, lack of positive

---

* In the 1950s and early 1960s strenuous efforts were made worldwide to develop an automated spinning mill; all these attempts to automate the card room sequence failed in that they did not become established as commercially viable systems (although some were successful technically). The main reason for failure was that the progress achieved in the productivity and versatility of the individual machines involved in the process overtook the labor-saving advantages of linking the machines, with the result the the use of semiautomatic features on the machines gave a greater advantage to the mill manager.[3]

Table 4-6. Details of Ring Spinning Mills for 20S and 30s Cotton—1950

Process	Hank	Speed	TPI	Efficiency (%)	Actual Production (lb/hr)	Waste Allowance (%)	Output Required (lb/wk)	Machines Required for Two Shifts
Raw fibre	—	—	—	—	—	—	109,000	—
Blowroom	16 oz/yd	—	—	75	342	5.5	103,400	2 lines
Cardroom								
Cards	0.14	26 in. doffer 12 rpm	—	86	12.2	5.5	98,000	108
Drawframes	0.10	150 ft/min	—	75	27.1	0.5	97,500	6×3×8
Slubbers	0.50	650 rpm	0.78	65	2.158	0.5	97,000	6×100
Inters	1.25	750 rpm	1.23	80	0.768	0.5	96,500	12×140
Ringroom	20	8500 rpm	18	92	0.043	0.5	96,000	74×400
	30	8500 rpm	22	94	0.024	0.5	96,000	127×420

Operatives Required per Shift (based on two shifts) and OHP

Process	Direct Operatives	Supervision and Ancillary	Remarks	OHP
Blowroom	2 attendants, 2 feeders, 1 head cotton-room man	1 baleman, ½ (1 wasteman and 1 baleman on days)	40-lb lap	0.547

Cardroom			
Cards	6 strippers and grinders, 3 tenters	12-in. × 36-in. cans	0.703
Drawframes	6 tenters	10-in. × 36-in. cans	0.547
Slubbers	6 tenters	1.75-lb doffed bobbins	0.625
Inters	12 tenters	1.5-lb doffed bobbins	1.015
Cardroom General	1 carder, 1 undercarder, 1 floor sweeper, ½ (1 cleaner on days)		0.273
Ringroom			
Count 20S	30 spinners, 10 doffers	6 breaks per 100 spindle hours, 6 patrols per hour at 30 secs. per frame. No overhead cleaning equipment, no suction floor cleaner. 600 bobbins per hour per doffer	
	1 overlooker, 2 jobbers, 2 tube sorters, 1 sorter, 1 sett carrier, 1 floor sweeper, ½ (1 oiler, 6 cleaners and 2 roller coverers on days)		4,100
		Count 20S total OHP	7,810
Count 30S	51 spinners, 10 doffers	As for count 20S ringroom	
	1 overlooker, 2 jobbers, 2 tube sorters, 1 sorter, 1 sett carrier, ½ (2 oilers, 8 cleaners and 3 roller covers on days), 2 floor sweepers		5,896
		Count 30S total OHP	9,606

Source: "Cotton and Allied Textiles—A Report on Present Performance and Future Prospects, Vol. 2 (Manchester, England: Textile Council, 1969).

Table 4-7. Details of Ring Spinning Mills for 20S and 30S Cotton—1968

Process	Hank	Speed	TPI	Efficiency (%)	Actual Production (lb/hr)	Waste Allowance (%)	Output Required (lb/wk)	Machines Required for Two Shifts
Raw fibre	—	—	—	—	—	—	105,000	—
Blowroom	16 oz/yd	—	—	67	335	5	99,500	2 lines
Cardroom	0.12	—	—	90	54.2	2	97,500	24
Cards	0.12	1,450 ft/min	—	75	216.5	0.5	97,000	6×2×1
Drawframes	1.10	1,000 rpm	1.16	72	1.12	0.5	96,500	12×96
Speedframes	20	13,000 rpm	18	92	0.0558	0.5	96,000	49×400
Ringroom	30	13,000 rpm	22	94	0.0366	0.5	96,000	88×400

Operatives Required per Shift (based on two shifts) and OHP

Process	Direct Operatives	Supervision and Ancillary	Remarks	OHP
Blowroom	1 attendant	2 balemen, feeders ½ (1 wasteman on days)	Bale pluckers and chute feed to cards	0.273

Cardroom			
Cards	1 attendant	36-in. × 42-in. cans	0.078
Drawframes	2 tenters	36-in. × 42-in. first passage,	0.156
		18-in. × 42-in. second passage	0.468
Speedframes	6 tenters	4.5-lb doffed bobbins	
Cardroom General	1 carder, 2 undercarders, 1 labourer		0.312
Ringroom			
Count 20S	9 spinners, 10 doffers	3.5 breaks per 100 spindle hours, 4 patrols per hour. Modern overhead cleaners and suction floor cleaner. 600 bobbins per hour per doffer	2.265
	1 overlooker, 1 jobber, 2 tube sorters, 1 sorter, 1 sett carrier, ½ (1 oiler, 1 roller coverer, 6 cleaners)	Count 20S total OHP	3.552
Count 30S	15 spinners, 10 doffers	As for count 20S ringroom	2.968
	1 overlooker, 2 jobbers, 2 tube sorters, 1 sorter, 1 sett carrier, ½ (2 oilers, 2 roller coverers, 8 cleaners)	Count 30S total OHP	4.255

Source: "Cotton and Allied Textiles—A Report on Present Performance and Future Prospects, Vol. 2 (Manchester, England: Textile Council, 1969).

Table 4-8. Comparison Among a Modern Open-end-spinning Mill, a Mill Built and Installed in 1963, and One Built and Installed in 1945[a]

	1945 Mill	1963 Mill	1973 Mill
Production			
Shifts operated	1	3	4
Hours per week	42.5	112.5	168
Production per hour (kg)	615	232	156
Production per week (kg)	26,122	26,122	26,122
Area (m^2)	7,500	3,250	2,100
Numbers and Types of Machinery Required			
Blowroom	6 scutchers	2 scutchers	1 chute-feed line
Cards	112	16	5
Drawframes (deliveries)	240	8	5
Speedframes (spindles)	2,520	384	—
Spinning frames (spindles/rotors)	36,000	9,504	1,700
Investment			
Machinery ($)			871,500
Ancillary equipment ($)			290,200
Building ($)			497,000
Total investment ($)		1,316,000	1,658,700
Investment per spindle/rotor ($)		(139)	976
Investment per employee ($)		(18,300)	51,834
Labor			
Total employees	176	72	32
Total production workers	168	64	24
H.O.K.	27.4	9.2	3.9
Direct workers per 1000 spindles rotors per shift	3.4	1.72	3.5
Labor cost per week (excluding administration) ($)	16,620[b]	7200[b] (2400)	3024
Labor cost per 100 kg ($)	63.63[b]	27.57[b] (9.19)	11.58

[a] Plant to produce 1,280,000 kg/annum of 34S-Nm (29.4-tex) yarn.
[b] These figures have been brought up to date to take account of present wage levels. The 1963 figures at 1963 rates are quoted in parentheses.
Source: F. A. Greenwood, "The Textile Mill of the Future," Proceedings, Textile Institute Conference, Asia and World Textiles. The Textile Institute, 1973.

Table 4-9. Comparison of a Conventional Worsted Spinning Sequence with a REPCO Self-Twist Sequence

	Machine Productivity m/min.	No. of Machs/ deliveries per operative	OHP	Kg. P.O.H.
Conventional Short-Staple Ring Spinning—24S Cotton Count				
Carding	77 lb/hr/m/c	33 Machs	0.039	1155
Drawing	244	6 Machs	0.130	350
Roving	28	437 Spdls.	0.190	239
Spinning	15.6	3327 Spdls.	0.616 Spinners	74
		7092 Spdls.	0.289 Doffers	157
Winding	914	28 Spdls.	1.544	29
Conventional Long-Staple Ring Spinning—2/23S Worsted Count (2/26S Nm)				
Drawing	115.2	4 Spdls.	0.300	152
Roving	53.2	76 Spdls.	0.338	135
Spinning	16.5	1082 Spdls.	1.177	39
Winding	900	29 Spdls.	0.915	50
Assembly winding	700	35 Spdls.	0.447	102
Twisting	41.8	519 Spdls.	0.505	90
Repco STT—11.5 Worsted Count Weaving Yarn (297 m/min REPCO) (STT Nm)				
Drawing	115.2	4 Machs	0.300	152
Roving	53.2	76 Spdls.	0.338	135
Spinning	297	15.5 Machs	0.648 Direct	70
			0.118 Indirect	385
Twisting	45	485 Spdls.	0.496	92
Winding	1000	16 Spdls.	0.769	59

Source: Technical Economy Department, Platt-Saco Lowell Ltd., Helmshore, England.

marketing approach, lack of financial stability, highly diversified markets, low production efficiencies, and high stocks.

As the result of an investigation by Viyella, the following policy of rationalization was adopted:

1. Reappraise the market situation and CEM's special place therein.
2. Rationalize the range of products and customers, giving the

large customer, who gave consistent and loyal support, reciprocal support.

3. Cater to Federation customers and gain the consequent savings that stem from the elimination of selling overheads, etc. Aim for vertical integration with its manifold benefits wherever appropriate.
4. Reappraise the re-equipment programs so that new machinery would more effectively improve quality and service to customers. Extend facilities for spinning synthetics.
5. Close down and liquidate unprofitable plants and excess capacity. Maintain multishift working on efficient equipment.
6. Simplify the management structure and cut down on the excessive headquarters staff. Ensure closer management links with the Viyella management committee to have the benefit of its expert services and resources.
7. Maintain a closer link with other yarn spinning and processing units of the Federation so that integration would continually take place on all fronts.
8. Carry out management education, so that managers would rediscover their commercial purpose and business acumen.

That this policy, when implemented, proved successful can be judged from Table 4–10.

Thus this rationalization of the overall operation had a marked impact both on total employment and on labor productivity, which clearly demonstrates that the more efficient use of existing plant can, in some instances, have as great an effect on employment as the adoption of new technology.

Changing Skills. Concomitant with changes in the nature of the technology used in the modern textile mill has been a broadening in the range of skills required by the mill's maintenance personnel. Perhaps the most notable example of this is the need to employ electronic engineers—or mechanical engineers with some training in electronics— to maintain the ever-increasing number of electronic monitoring and control devices incorporated in modern machinery.* A second manpower-related factor resulting from the increased use

---

* In many instances the onus has been on the machinery-producing companies to train mill operatives and overlookers in the correct usage of their new machines. Certainly machine builders have striven to design electronic systems to make their maintenance as simple and as speedy as possible.

Table 4-10. CEM: Before and After

	Before (1964)	After (1966)
Number of Spinning Mills	14	7
Number of Operating Subsidiaries	8	—
Weight Produced Per Year (million lb)	28	25
Spindles installed	404,000	221,000
Production per Spindle Year (lb)	69.91	113.0
Average Running Time per Spindle (hr/wk)	56.9	110.5
Capital Employed (millions)	8.0	3.5
Number Employed	3,983	2,318
Production per Employee per Year (lb)	7,020	9,920
Stocks of Yarn	£754,000	£196,000
Production Floor Area (million ft^2)	4.3	2.0
Management Accounting Staff	3	7
Centralized Staff	43	17
Customers	735	120
Bank Balance	−£988,000	+£200,000

of electronic devices that reduce the possibilities for manual error and make it easier to control textile machines is that the skill content of the operative's job, in some instances, has been reduced.

Government Intervention. The bulk of the textile industry in the United Kingdom is situated in the development areas where government assistance is available to establish new factories. This has probably had some effect on maintaining employment in this sector. However, without resorting to protectionism, it is difficult to see what the government can do to halt the current marked decline in the textile industry brought about partly as a result of the current world recession and partly because of cheap imports.

Summary

Technical change, often of a radical nature, has been an increasingly important component in competition in textile machinery during the post-war years. This has imposed the requirement on machine builders to employ development personnel with higher qualifications than before. Certainly most of the radical innovations in this area have been associated with the presence of qualified engineers and sometimes scientists and with a formal, systematic research and development effort. Additionally, technological change in textile machinery has played a significant part in reducing the manpower component of textile production. Rationalization and concentration in the textile industry with the formation of large, well-equiped, integrated production units has also resulted in some reduction of employment in this sector. Finally, the use of new technologies in textile production has led textile firms to employ personnel with a greater range of skills.

5 Technological Innovation International Trade Patterns

Regina Kelly

The role of technological innovation in explaining the level and composition of a country's exports has recently gained the forefront in international trade theory. Most efforts in the past concentrated on the comparative cost and factor proportions theories of comparative advantage as the central explanations of why nations trade and why they benefit from trade. Put simply, the classical theories state that countries trade because of differences in their relative endowment of the factors of production (Heckescher-Ohlin theory) or comparative productivity (Ricardian theory).

The factor proportions and comparative productivity explanations have been challenged, or at least augmented, by technology proponents in two principal variations. The first considers technology to be a form of "human" capital, usually evidenced in a highly skilled labor force. This is the explanation Leontief used to reconcile the apparent contradiction in his finding that U.S. exports are labor rather than capital-intensive. Introducing technology in this fashion, however, produces a rather static picture of the determinants of trade as its rests on the assumption of a standard world technology. Relative factor endowments, which may change only slowly over time, determine the direction of trade.

The second variant, which is more dynamic, is commonly associated with the name of Ray Vernon and the "product life cycle" theory. In contrast to a standard world technology, technical knowledge is assumed to be neither a costless nor a universally available good, and its possession creates a transitory advantage for the exporting country. Products that are new, highly sophisticated, or technologically advanced are highly differentiated. Since there are few substitutes and few, in any, competitors, theory suggests that such trade is not very price-sensitive, competing on the basis of its "uniqueness" instead. In this scenario, technology is viewed as continuously changing, and these changes alter trade patterns. Innovation leads to exports, and only as technology diffuses to other countries does the pattern of world trade revert to one based on factory proportions and, hence, on relative prices.

The purpose of this paper is not to see how well rival hypotheses of the determinants of trade perform. Empirical testing of these and other theories of trade not mentioned here have shown that each may have substantial explanatory power, especially in regard to a limited number of countries or products, but none has sufficiently commanding power to dispose conclusively of rival claimants.

Rather, this paper presents an initial examination of recent movements in trade flows to see whether a technology factor is readily identifiable and, if so, whether it conforms to generally held expectations. In the words of Harry Johnson, "Every blind man who touches a part of the elephant learns some truth about it—but not the whole truth; and only the rare unfortunate is unlucky enough to be caught generalizing about the elephant from an unrepresentative hand-hold on the tip of its tail." It is hoped that this paper is more, rather than less, representative.

Methodology

For purposes of this examination, OECD exports of manufactures were divided into two groups, "technology-intensive" and "nontechnology-intensive" manufactures, for the years 1968, 1971, and 1974. The base year 1968 was chosen to minimize any possible distortion from differential rates of economic growth in the United States and abroad, while data for 1974 are the most recent available in a form appropriate for this task. In addition, this time frame seems appropriate since it represents a period of increased government attention to science and technology policy and considerable movement among the major OECD countries in their relative commitment of resources to research and development.

The products termed "technology-intensive" in this paper may be thought of as those that lend themselves most readily to innovation.

Briefly, the products were identified as technology-intensive or nontechnology-intensive by examining U.S. data on applied research and development by product field in the manufacturing sector and relating these expenditures to product shipments.[1] Product rather than industry data were used to correct the problem of industry diversification of research and development expenditures to many different product classes.

As shown in Table 5-1, the average research intensity for all U.S. manufacturing during 1968–1970 was 2.36%—that is, 2.36¢ of research and development expense per dollar of sales. Those products of above average research intensity were classified as technology-intensive. A concordance was then developed to match the U.S. product codes with the Standard International Trade Classification (SITC).

Aircraft, computers, and electronics rise to the top of the technology-intensive category, followed by such products as engines, turbines, and petrochemicals. The nontechnology-intensive category is comprised of such products as automobiles, construction machinery, semimanufactures, and textiles. Overall, the results appear reasonable to that which might be expected a priori.

This methodology results in a definition of technology-intensive trade that is considerably narrower in scope, and hence conceptually preferable, than that employed in previous research, but it must be critically viewed in at least three respects: 1) it uses quantifiable research-and-development intensity as a proxy for technology intensity; 2) though its product group identification is the most detailed that present statistics allow, it is still too aggregative to prevent product misclassification from occurring; and 3) it is based on the product distribution of research and development in the United States rather than in the OECD and may thus be distorting when applied to other countries.

The distortion introduced by these difficulties, however, does not appear to be of a critical magnitude. Tests have shown that research-and-development intensity correlates very highly with all other indicators of technology intensity.[2] The product detail, while far from ideal, is the most specific possible and yields objective results that are close or identical to most subjective judgments.

Some distortion is certainly introduced by the use of product research-and-development ratios in the United States to define technology-intensive products for the OECD as a whole. Comparisons based on available data regarding research-and-development distributions in other OECD countries, though, tend to indicate a

Table 5-1. Description of Product Classes by Technology Classification, 1968–1970

U.S. SIC Code	Position Description	Research Intensity Ratio (%)[a]
Excluded[b]		
*1925	Guided missiles and spacecraft	84.52
Technology-Intensive		
372	Aircraft and parts	12.41
357	Office, computing, and accounting machines	11.61
361–362, 366–367	Electric transmission and distribution equipment; electrical industrial apparatus; communication equipment and electronic components	11.01
383–7	Optical and medical instruments, photos, watches	9.44
283	Drugs and medicines	6.94
282	Plastic materials and synthetics	5.62
351	Engines and turbines	4.76
287	Agricultural chemicals	4.63
*19 less 1925	Ordnance, except guided missiles	3.64
381–382	Professional, scientific, and measuring instruments	3.17
381	Industrial chemicals	2.78
365	Radio and TV receiving equipment	2.57
Nontechnology-Intensive		
352	Farm machinery and equipment	2.34
371	Motor vehicles and equipment	2.15
363–364, 369	Other electrical equipment and supplies	
353	Construction, mining and related machinery	1.90
284–286, 289	Other chemicals	1.76
34	Fabricated metal products	1.48
30	Rubber and plastic products, n.e.c.	1.20
354	Metalworking machinery and equipment	1.17
373–375, 379	Other transportation equipment	1.14
*29	Petroleum and coal products	1.11
355–356, 358–359	Other nonelectrical machinery	1.06
*21, 23–27, 31, 39	Other manufactures, n.e.c.	1.02
32	Stone, clay, and glass products	0.90
333–336, 3392	Nonferrous metals and products	0.52
331–332, 3391, 3399	Ferrous metals and products	0.42
22	Textile mill products	0.28
*20	Food and kindred products	0.21
Total manufacturing		2.36

[a] The ratio of applied research-and-development funds by product field to shipments by product class.

[b] SIC 1925 (guided missiles and space craft) was excluded from the calculation of the average intensity ratio of U.S. manufacturing because of its extremely high-intensity ratio—almost seven times as high as the nearest product class—and its limited importance in U.S. trade flows.

* Trade in SIC 19, 20, 21, and 29 is not discussed in this paper because they are not included in the Standard International Trade Classification definition of manufactures SITC 5.

high degree of correlation with the U.S. pattern (around 0.7, significant at the 5% level). It would appear that products lending themselves to research and development and to innovation in the United States are in large part the same group of products lending themselves to innovation in other OECD countries. Thus there is some evidence for relying upon a global definition of technology-intensive products.[3]

Initial Expectations

Five countries—the United States, the United Kingdom, Germany, France, and Japan—account for roughly 95% of OECD expenditures for research and development. These five countries, however, account for a considerably smaller relative proportion of OECD income and manufactures exports. Thus one might hypothesize that these five countries have a comparative advantage in technology-intensive products and, therefore, expect to find a number of characteristics that would distinguish their trade from that of the remaining OECD countries:

1. A relatively higher proportion of their manufactures exports might be expected to be technology-intensive products.
2. They should command a higher share of total trade in technology-intensive products than in nontechnology-intensive products.
3. Their technology-intensive exports might be expected to grow faster than those of the remaining OECD countries and faster than their own nontechnology-intensive exports.

OECD Exports of Technology-Intensive Products

In examining how well the data support these hypotheses, it might first be supposed that OECD exports of technology-intensive products would experience a sharply higher growth rate than OECD exports of nontechnology-intensive products. Demand for technology-intensive products are presumed to be locally produced in a much smaller number of countries than are nontechnology-intensive products, common wisdom suggests that they would be the major growth factor in the export trade of the developed countries. Such a dramaitc distinction, however, is not borne out by the data.

46 Overview: Economic and Social Aspects of Innovation

From 1968 to 1974, total OECD technology-intensive exports grew at a compound annual rate of 22.4%, contrasted with a compound annual growth rate of 20.9% for nontechnology-intensive products. This difference is not insignificant, for it implies that technology-intensive products were gradually accounting for an increasing share of total OECD exports of manufactures. Over the entire 6-yr period, technology-intensive exports grew about 10% more rapidly than nontechnology-intensive exports. As a result, technology-intensive products rose marginally from 28.4% of total OECD exports of manufactures in 1968 to 29.9% in 1974. Although not insignificant, these figures nevertheless are considerably less dramatic than has been frequently supposed.

Export Composition of the Research-Rich Countries

What is perhaps more suprising, however, is the export behavior of the five "research-rich" countries that account for 95% of OECD research and development expenditures: France, Germany, Japan, the United Kingdom, and the United States. At first glance, the technology factor appears to be considerably more significant for these countries than for the OECD as a whole (see Table 5–2). In 1968, technology-intensive products comprised 31% of the five countries' total exports of manufactures, compared to 24% for the rest of the OECD countries. Moreover, these countries accounted for 72% of OECD technology-intensive exports, compared to 65% of nontechnology-intensive exports.

Upon closer examination, however, these apparent differences are attributable to one country—the United States. As shown in Table 5–2 40% of U.S. exports of manufactures during this period were technology-intensive products, a far higher proportion than for any other country. There were no significant differences among the other four research-rich countries. Technology-intensive products comprised roughly 26% of the manufactures exports of each of the four, despite the fact that the four countries expend differing proportions of their GNP on research and development.

Even more surprisingly, there was no significant difference between these four countries, taken either individually or as a group, whn compared with the remaining OECD countries. Technology-intensive products accounted for 24% of the manufactured exports of the remaining OECD countries, only marginally less than the figure for the research-rich countries other than the United States.

Table 5-2. OECD Exports of Manufactures, 1968 and 1974
(billions of current $)

	Total OECD	U.S.	Japan	France	Germany	U.K.	Five Research-Rich	Rest of OECD
			Composition of exports ($)					
1968								
Technology-Intensive	34.2	9.6	3.2	2.4	5.8	3.5	24.5	9.8
Nontechnology-Intensive	86.4	14.1	9.0	6.9	16.4	9.2	55.7	30.7
Total manufactures	120.6	23.7	12.2	9.4	22.3	12.7	80.2	40.5
1974								
Technology-Intensive	115.1	26.6	13.2	8.6	22.0	10.2	80.4	34.7
Nontechnology-Intensive	270.0	35.9	35.2	23.5	54.2	21.7	170.4	99.4
Total Manufactures	385.1	62.4	48.3	32.1	76.2	31.9	250.9	134.2
			Composition of exports (%)					
1968								
Technology-Intensive	28.4	40.5	26.0	26.0	26.2	27.2	30.5	24.2
Nontechnology-Intensive	71.6	59.5	74.0	74.0	73.8	72.8	69.5	75.8
1974								
Technology-Intensive	29.9	42.5	27.2	26.7	28.9	31.8	32.1	25.8
Nontechnology-Intensive	70.1	57.5	72.8	73.3	71.1	68.2	67.5	74.2

Source: OECD, *Trade by Commodities,* Series C (1968 and 1974), See Note 4 for data comments.

48 Overview: Economic and Social Aspects of Innovation

Thus the United States stands out as the only country with an unusual concentration of its manufactured goods exports in the technology-intensive category. Such a concentration is an indicator of comparative advantage; and although such indications were expected for the other research-rich countries, they did not materialize. In fact, the results for the countries other than the United States were not different from those that would have been expected if the relative research-and-development expenditures of each of the countries were identical to every other country.

Some distinction among the research-rich countries does appear, however, when allowances are made for the differences in the research-intensity of each of the individual product classes that comprise the technology- and nontechnology-intensive groups. By weighing the research-intensity ratio of each product class by its proportion of a country's total manufacturing exports, it is possible to derive a single number that reflects the technology mix of that country's exports.

As previously mentioned, the average research-intensity for all U.S. manufacturing in 1968 was 2.36%; the average technology-intensity of U.S. manufactures exports in 1968 was 4.23%—almost 80% higher than domestic production. Applying this methodology to the manufactures exports of the other OECD countries shows the following relative export technology-intensities: the United Kingdom, 3%; France and Germany, 2.7%; rest of OECD, 2.5%; and Japan, 2.36%.

These export technology-intensity ratios are closely correlated to the relative commitment of resources to research and development on the part of each of the countries during this period, as is shown in Figure 5–1. Thus it would appear that the more a country expends on research and development relative to the size of its GNP, the more technologically sophisticated are its exports.

Export Competitiveness of the Research-Rich Countries

The technology factor appears extremely weak when sought in terms of the relative competitiveness of the countries. "Competitiveness" is defined here as a country's share of total OECD exports for each product class and for the technology groups.

Again, the United States is the exception. As measured by the Spearman rank coefficient, the correlation between the United States share of total OECD exports of a product class and the research-

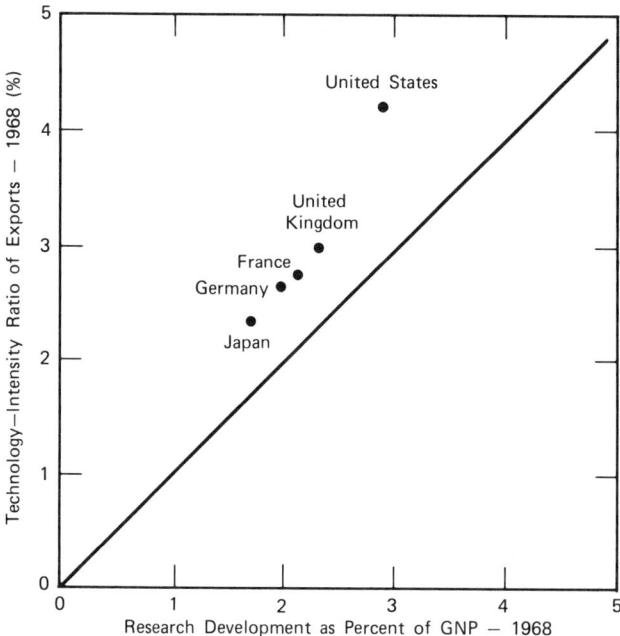

Figure 5-1. Technology intensity of manufactured exports compared with relative expenditures on research and development.

intensity of a product class is 0.68, significant at the 1% level. This association is most clearly illustrated in Table 5-3. In 1968 the United States had a 28% share of the technology-intensive exports of the OECD countries, but only 16% of the market for nontechnology-intensive products. Thus the U.S. share of technology-intensive OECD exports was 75% higher than in nontechnology-intensive exports.

The United States, however, was the only country for which this distinction existed. Each of the other OECD countries had a larger or equal share of nontechnology-intensive exports than of technology-intensive exports. Statistical testing for each country, moreover, showed no significant relationship between share performance and the research-intensity of the product.

This result, it must be noted, was not predicated on the fact that for every winner there must be a loser (i.e., that a relatively large U.S. share in technology-intensive products would mask the differences that really exist for the other countries). For example, if U.S. exports are excluded, the four research-rich countries have only a

Table 5-3. Share of OECD Exports of Manufactures, 1968 and 1974

	Share of Total OECD Exports (%)		
	Technology-Intensive	Nontechnology-Intensive	Total Manufactures
1968 Total OECD	100.0	100.0	100.0
United States	28.0	16.3	19.6
Japan	9.3	10.4	10.1
France	7.1	8.0	7.8
Germany	17.0	19.0	18.5
United Kingdom	10.1	10.7	10.5
Rest of OECD	28.5	35.5	33.5
1974 Total OECD (value)	100.0	100.0	100.0
United States	23.1	13.3	16.2
Japan	11.4	13.0	12.5
France	7.4	8.7	8.2
Germany	19.1	20.1	19.8
United Kingdom	8.8	8.1	8.3
Rest of OECD	30.1	36.9	34.9
1974 Total OECD (volume)[a]	100.0	100.0	100.0
United States	26.7	15.7	19.0
Japan	10.0	11.6	11.1
France	7.7	9.2	8.7
Germany	17.0	18.2	17.9
United Kingdom	9.0	8.4	8.6
Rest of OECD	29.6	36.9	34.7

[a] Current dollar export value deflated by dollar unit value indices for exports of manufactures (1968=100).
Source: OECD, Trade by Commodities, Series C (1968, 1971, and 1974). See Notes 4 and 5 for data comments.

marginally higher share of technology-intensive than nontechnology-products, 60.4% vs 57.6%. Even this difference is based solely on their low share of "miscellaneous manufactures."

Export Growth of the Research-Rich Countries

When the growth rates of the individual countries are examined, there is no statistically significant difference between the distribution of the technology-intensive and nontechnology-intensive export growth. Those countries whose exports of technology-intensive products grew rapidly also exhibited better than average growth in their nontechnology-intensive exports and vice versa.

This result hold true whether 1974 exports are examined on the basis of current dollars or constant dollars, that is, correcting for changes in export prices resulting from exchange rate movements and differential rates of domestic inflation. However, when exports are viewed in real terms and the period of 1968–1974 is divided into two subperiods, two interesting indications of a distinction between the performance of the technology-intensive and nontechnology-intensive export groups appear.

The first, not surprisingly, is in the exports of the United States. During 1968–1971, U.S. exports of technology-intensive products grew at a compound annual rate of 7.1%, the slowest rate of growth among the OECD countries (see Table 5–4). In contrast to the growth of U.S. nontechnology-intensive exports, however, this performance was strong; U.S. nontechnology-intensive exports grew at a compound annual rate of only 2.9% compared with the average OECD growth rate of 10.4%.

These years represent a period of an overvalued dolar, and subsequent to 1971 a remarkable turnaround is evident. U.S. technology-intensive exports recovered in the 1971–1974 period and grew somewhat faster than the OECD average. Even more dramatic, however, was the recovery in U.S. exports of nontechnology-intensive products, which grew at a compound annual rate of 15.8%—more than 6 percentage points faster than the exports of any other major OECD country.

The United States gains in nontechnology-intensive products in the latter period are in contrast with the case of the United Kingdom, which is the second instance of a distinct difference in export performance by technological categorization. The export performance of the United Kingdom during 1968–1971, can be characterized as generally weak. During 1971–1974, the relative position of technology-intensive exports declined somewhat further in comparison with the rest of the OECD; the position of the United Kingdom in nontechnology-intensive exports, however, declined precipitously —with exports of nontechnology-intensive products growing at less than half their previous rate.

Relative Technological Position in 1974

In terms of shifts between the countries, the technology-intensity ratios for each country's manufactures exports provide one of the

Table 5-4. Real Growth of OECD Exports of Manufactures[a] (%)

	Real Growth of Exports (compound annual rate)		
	Technology-Intensive	Nontechnology-Intensive	Total Manufactures
1968–1974 Total OECD	11.6	9.9	10.4
United States	10.7	9.2	9.8
Japan	13.0	11.9	12.2
France	12.9	12.3	12.4
Germany	11.6	9.1	9.8
United Kingdom	9.6	5.6	6.8
Rest of OECD	12.2	10.6	11.0
1968–1971 Total OECD	10.2	10.4	10.3
United States	7.1	2.9	4.6
Japan	18.9	17.0	17.5
France	11.2	15.4	14.3
Germany	9.3	8.7	8.9
United Kingdom	9.3	7.9	8.3
Rest of OECD	10.6	11.9	11.6
1971–1974 Total OECD	13.0	9.4	10.4
United States	14.4	15.8	15.2
Japan	7.5	7.0	7.1
France	14.7	9.3	10.6
Germany	13.9	9.5	10.7
United Kingdom	9.8	3.4	5.2
Rest of OECD	13.9	9.3	10.4

[a] Current dollar value deflated by dollar unit value indices for exports of manufactures (1968=100).
Source: OECD, Trade by Commodities, Series C (1968, 1971, and 1974). See Notes 4 and 5 for data comments.

most interesting pictures since they allow for relative changes within the technology groups from lower to higher research-intensive products, as well as reflecting a change in the distribution between the two technology groups. By 1974 Japan had raised its technology-intensity ratio to the average of the rest of OECD, or 2.5%; the United States and the United Kingdom also increased their intensity ratios, to 4.4 and 3.2%, respectively. The other OECD countries remained virtually unchanged.

An increase in the technology-intensity ratio of a country's exports does not necessarily imply an increase in a country's ability to compete in technology-intensive products, however. For example, between 1968 and 1974, Japan increased its share of OECD exports in both technology groups; and the distribution of the share gain was such that by 1974 a larger proportion of Japanese exports was concentrated in relatively higher-technology products. In contrast, the United States and the United Kingdom experienced a share loss in both technology groups for the period as a whole. The loss was relatively greater, however, in nontechnology-intensive than in technology-intensive products. Thus the increase in the importance of higher-technology products in the export composition of these two countries more appropriately reflects their relative inability to compete in lower-technology products.

No major shifts took place during the 1968–1974 period between the relative position of the research-intensive countries and the rest of the OECD, but there were some changes among the major OECD countries. Despite some share loss, the United States retained its position as the world's largest exporter of technology-intensive products. This also held true for total manufactures in terms of constant dollars, but not in current dollars. The most significant change, however, was that the United Kingdom fell from the third largest exporter of manufactures in 1968 to the fifth largest in 1974. By 1971 Japan had surpassed the United Kingdom in exports of both technology- and nontechnology-intensive products, and by 1974 France had surpassed the United Kingdom in nontechnology-intensive exports.

Conclusions

The technology factor does not appear to distinguish itself in this measure or in the trade characteristics of the OECD countries. Only the United States stands out among the OECD countries in the high concentration of its exports in technology-intensive products and in the strong positive relationship between the technology content of its exports and its competitive performance in international trade.

Several factors may be suggested for the lack of more evidence of strong trade–technology relationships. First, of course, there may be blame in the admitted data weaknesses of the approach in this paper. One possible explanation for less dramatic results than might have been expected is the application of research-intensity ratios derived from U.S. domestic production to the trade of other countries. This assumes that the relative research-and-development position of given products is similar in different countries and that the mix of goods within each individual product group is also comparable. Indications are, however, that research-and-development expenditures are similar enough across countries, and the trade relationships are so weak and dispersed, that it must be questioned whether the use of individual country definitions of technology intensity would seriously change the results obtained here. Instead, a more likely source of error may be the unavoidable misspecification of some lower-technology products as technology-intensive and vice versa.

Another possibility could be that the technology factor has been sufficiently diffused through the transfer process that only the United States, relatively speaking, derives from its research efforts a comparative advantage that is demonstrable in trade. Overall, technology by itself was inadequate to explain the primary trade characteristics of the countries. At this level of disaggregation, the introduction of a technology factor was no guarantee of a virtual monopoly by a few countries of trade in a particular product.

In only a few instances—such as the United States in aircraft, Japan in consumer electronics, and, to a lesser extent, Germany in petrochemicals and industrial chemicals—could a virtual monopoly by one or a small number of countries be seen in the various technology-intensive categories. Moreover, one can think of other factors, such as economies of scale in aircraft and price in the case of televisions that might be factors bearing on the disproportionate share.

On the whole, those countries that performed well in their nontechnology-intensive exports also experienced good growth in their technology-intensive exports. This result suggests that the two technology export groups shared factors in common that accounted for their competitiveness. There were tantalizing suggestions, however, of technology working at the margin.

Technology-intensive products are assumed to compete principally on the basis of factors other than price, but on preliminary examination the data actually showed some interesting indications of price sensitivity in the technology-intensive export category. This seemed to show up in the cases of the United States and the United Kingdom, and to some extent Japan. During periods of poor price performance, the relative growth of technology-intensive exports was reduced from previous and subsequent experience, though it was not nearly as affected as the nontechnology-intensive group.

Nonetheless, the fact that the technology-intensive category seems to be reacting to price somewhat holds some interesting inferences for the United States. First, some observers, noting the declining trade balance in technology-intensive products in the late 1960s and early 1970s, have expressed concern over the declining ability of the United States to compete in these products. The subsequent strong growth in the U.S. technology-intensive exports after the currency realignments, however, suggests that they may have been adversely affected earlier by an overvalued dollar. Second, the overvalued dollar for much of this period may have resulted in an exaggeration of the apparent U.S. reliance on innovation in its manufactures export trade.

For the major OECD countries individually, as well as for the OECD as a whole, there has been a gradual shift in the composition of manufactures exports toward the technology-intensive products, but this movement has been slow enough to make it clear that technology is only one factor out of many influencing trade flows. Many of the fastest growing product classes were nontechnology-intensive, such as ships, motorcycles, rubber products, and construction machinery. Imbalances in supply and demand, such as resulted in the export boom of iron and steel mill products in 1974, the liberalization of trade barriers, differential rates of growth in income, labor costs, productivity, inflation, or exchange rate changes may well act to swamp the effects of the technology factor in trade, especially in the short term. Surprisingly, in fact, rather than creating dramatic changes in export patterns, the technology-intensive products appear in some ways to be acting as a source of stability in the face of more volatile changes in nontechnology-intensive exports.

Stark differences in international competitiveness are difficult to tie directly to the technology content of products. On the basis of this analysis, only the United States may be said to have a relatively heavy export reliance on technology-intensive products. The technological comparative advantage this implies, however, may be misleading. The high degree of correlation between economic variables make it impossible in an examination of this type to precisely specify the relationship between underlying factors.

For instance, in the case of the United States, considerable research-and-development activity is strongly correlated with the comparatively intensive use of high, skill level manpower in general. As Ray Vernon has stressed, for some policy purposes the distinction between whether research and development effort or the pattern of manpower use in general is the prime factor influencing export performance can be crucial. Thus it is appropriate to emphasize that, although technology may be important, it is only one factor influencing trade flows.

On the more positive side, it should be stated that trade flows do not necessarily reflect shifts in final demand. Recent trade trends indicate that OECD technology-intensive exports (and hence world imports) are growing only somewhat more rapidly than nontechnology-intensive exports. Other influences, such as trade liberalization or growth in the multinational corporations, however, may be affecting the distribution of export growth between technology-intensive and nontechnology-intensive exports.

In addition, it is possible to speculate that innovation may become a more important factor in future OECD trade flows if the less-developed countries (LDCs) growing aspirations to increase the capabilities of their manufacturing sector are realized. The reliance of the various OECD countries on the LDCs as a market for their manufactures exports varies considerably, from roughly half for Japan to approximately one-fifth for continental Europe. Even at present the LDCs are more important as a market for the technology-intensive exports of the OECD countries than for their nontechnology products (with the exception of Japanese exports). This pattern may be accentuated in the future if the LCDs are to specialize in the manufacture of products in which they have a comparative advantage. To the extent that innovation creates a demand for products that the LCDs cannot manufacture efficiently for themselves, then the demand for the exports of countries that are innovators may exceed that of non-innovators.

Finally, innovation is a direct means of satisfying human wants and needs by increasing the variety of new products and the application of old ones. In addition to product-differentiating change within a product line, major innovations may also occur in process technology, which could apply outside of this line to, say, the nontechnology-intensive products. Innovation plays an important indirect role through increases in the overall factor productivity of human and natural resources. It may well be that the primary influence of technological innovation on international trade flows is via this latter contribution, with its effect on economic growth and relative prices. As such, the measure of technology, via research-and-development expenditures may not be a complete measure of the overall importance of the influence of technology on the exports of the developed countries.

6 Human and Social Aspects of Technological Change

Aubrey Kagan

The previous analyses focused on the effects of innovation on employement quality and levels; however, it is now time to turn to the human and social effects of innovation. This analysis focuses on the work environment in Sweden.

There is no doubt that Sweden has a high standard of living. It has been high for a decade or so and rose in the post-war years from a relatively low level to the present "highest in the world."

Probably the single most important cause for this was the export boom during these same years. Sweden had a well-equiped and organized industrial force before the war. It was able to capture orders in a seller's market immediately after World War II and retain them later by the high quality of goods produced by Swedish machine and energy-intensive methods.

The unusually broad distribution of the good things that can be bought with the profits of full employment and adequate markets is largely a result of the approach of the Social Democratic government in power for 44 years until 1976. The Social Democrats still have more votes than any other single party.

There are no slums in Sweden. No one lacks food, shelter, clothing, or high-quality health services. Education is free and the university level board and lodging can be supported by interest-free loans. The transportation and communication systems are well developed. Nearly everyone can be communicated with by telephone. There are five TV sets for every 10 persons. Newspaper coverage is extensive and TV debate frequent. People are well educated and have ready access to information about each other and to government business. In spite of high taxation, purchasing power is high. Because of high taxation, the net income of the least skillful workers is not much less than about half that of the most skillful. Sickness and unemployment have little influence on purchasing power.

Sweden has a tradition going back to the nineteenth century of being avant garde in its social policies directed toward humanizing work. Many of these policies relate by practice or by law to innovations in industrial relations. Some recent legislation regarding the humanization of work and some of the activities planned for the future are briefly mentioned.

The Workers Protection Act of 1949 and the amending ordinances that came into force in 1974 detailed provisions for safety, hygiene, and the avoidance of physical hazards. One of the innovations in 1974 enabled the "safety delegate"—a worker elected by the unions—to order suspension of work he considered unsafe, posing "immediate and serious danger for life or health," pending a decision by the Labor Inspectorate. The safety delegate would not be held liable for any damage resulting from his action.

The proposed act on Work Environment scheduled to supersede this law in 1978 is very different. It puts the onus on employers to see that work conditions are adapted to the workers physical and mental needs and capacities. It emphasizes psychosocial factors, but it does not specify what these are or might be. Two government agencies are given wide discretion to issue generally applicable rules and also prescriptions addressed to a specific firm. These agencies can decide what should be done by whom. The role of the courts is limited to examining whether the agency had formal competence to take the decision in question—not likely to be in doubt—and adjudicating on any proposals for imprisonment.

The act on Codetermination at Work that came into force on January 1, 1977, is also an "open frame" directive to the employers. They are bound always, or in some cases only on the demand of the employees, to negotiate with union representatives (often several because more than one union is involved) on all changes that may affect working conditions before they are enacted. Initial discussions must be with local union representatives. If there are none, or if representatives of different unions disagree or are not satisfied, the central union representatives must negotiate with the employer. The final decision rests with the employer. If he disagrees with the union representatives, negotiations may be drawn out. On the other hand, the employer can be taken to a labor court if he refuses to negotiate.

The court consists of a professional judge with a casting vote, one or two trade union representatives, one or two employer's representatives, and one or two lay people. The majority decision is final and there is no appeal.

The Protection of Employment Act, which came into force in 1974, virtually prevents an employee from being dismissed if he has been employed for more than 11 months. The labor court can be asked to judge whether "reasonable grounds" for dismissal existed. This act also provides for redundancy of workers.

The Act to Protect the Elderly and Handicapped of 1974 requires employers to consult a local government agency about measures to take to ameliorate the situation of employed elderly and handicapped people and to recruit such persons. The government agency can prescribe specific measures to be taken by any employer the agency chooses. In case of disobedience the matter goes to a central government agency, and if the employer concerned does not obey the prescriptions issued by that agency, the latter can prohibt him from employing any persons other than those sent to him or approved by the local employment agency. If the employer fails to comply with such an obligation, he may be fined or, less likely, put in prison. Again this act is an open frame law activated, judged, and executed by government agencies.

The Flexible Retirement Law of 1976 permits people to semiretire or retire at age 60 on lower retirement pay, at age 65 on normal retirement pay, or up till 70 on higher retirement pay. Retirement pay is geared to the cost of living.

An Act for Equal Opportunities between Men and Women in all areas of work is in preparation. An equal opportunities law applicable to government agencies only has been in effect since 1976. Government agencies are obliged to take steps to promote equality between men and women. This again is an open frame law. Each agency reports its activities yearly by the budget ministry. Another approach to ensure employment equality between men and women is the rule that an employer cannot get government subsidies for a new plant unless he employs at least 40% women and at least 40% men. Another project encourages women to take up traditionally male jobs. Employers are offered a compensation of $\simeq$$3 per hour for 6 months for each recruited woman put into a traditionally male job. Another approach to encourage women to work is a law that enables a father or mother to take paid maternity leave or to share it. Day care in nurseries for preschool children also supports equal opportunities for women to work.

Many of the measures to protect workers' interests described here have been initiated and practiced by avant garde industries in advance of the laws. Two companies that are ahead of the present laws are Volvo and Saab. They have experimented to change motor engine assembly from a moving belt with each worker doing repetitive small operations to small teams of workers assembling a whole engine in a bay, with the right to choose how they do it.

At Volvo they believe that people, not machines, are the real basis for the spectacular growth of industry during the twentieth century. The management at Volvo believe that economies of scale have subtle limits. They are convinced that people do not want to be subservient to machines and the dominance of impersonal processes had resulted in high absenteeism, apathy, antagonism, and possibly malicious mischief.

The Volvo factories at Kalmar and Skövde may no longer be considered "experiments." The Kalmar plant assembles cars in working groups of about 20. At Skövde the product is an engine, but the principle of group production remains the same.

There is no doubt that government, trade unions, industry, civil servants, and workers themselves have cooperated to a large extent in initiating and carrying out all these measures. What motivates them is hard to say. Ostensibly, and I think in reality to a large extent, they are directed to providing equal shares in a better quality of life for all people. But altruism is not usually enough. At the national level there is justifiable pride that a small nation can set examples to the world and there is, with good reason, international surprise and acclaim.

Any government that seeks popularity must please the electorate and, in Sweden, the powerful, well-organized trade unions as well. A government that takes $10 for every $4 that the average person spends must give something back for this. Traditionally the return has been in the form of services and more gross salary for the workers and higher taxes for the higher payed or apparently privileged.

If trade unions are to increase their status, they must appear to have been instrumental in achieving gains. Industry, whether private, state, or mixed, has had an excellent record in management and enterprise. The power of the government and the trade unions, which has risen on the basis of good industrial management, diminishes that of industry. Traditionally, the industrialists are the "capitalist bogeymen" of the government and the unions. In fact, they cooperate well. With good times, high profits, and high taxes', it is reasonable for industry to support the desire to improve conditions for the majority on the one hand and the underprivileged minorities on the other. A large organization retains some of its power when it works with the community and not against it. Productivity is not so important when there is plenty for all.

After 44 years of successful social democratic rule, the civil service, not surprisingly, has become imbued with the social democratic principles and philosophy. "The way to get ahead is to be a Social Democrat." Few, whatever their beliefs, speak against the party line in public. It would be a little like speaking against the monarchy toward the end of Queen Victoria's reign in Britain.

The mass media debates innovative ideas and burning issues daily. It seems to be biased toward the left, blaming capitalism for all evils. In the debate on equal pay for all workers, for example, nobody took the view that although society might prefer to do its own cleaning if cleaners asked for higher pay no one would do his own surgery even if surgeons asked for higher pay. In fact, a surgeon took the view publicly that cleaning floors was just as important as his work. In the TV debate on prostitution, no one denied the untenable and extraordinary contention that it was the fault of landlords who charged high rents for rooms of assignation.

The Swedish people themselves are accustomed to a high standard of living, and are more interested in conditions of work that satisfy needs of belonging, status, and self-expression than in more pay, unless someone else has gotten a raise. A high proportion, but by no means all, want a higher share in saying how they do their own job. Fewer want a say in how the factory should be run as a whole. Many, but not all, want more variety, no short repetitive jobs paced by a machine or an assembly line, a feeling that they have contributed to what has been produced, and good two-way relations with colleagues and supervisors. They support directly the union and government that goes out of its way to get them these things. Workers generally support the Social Democrat Party financially, because unless they state their wish to the contrary each year, a proportion of their union dues goes into the Social Democratic Party's coffers. This is convenient for those who really wish to contribute and awkward for those who don't.

Innovative laws must be judged on the basis of improving well being and health. Their purpose has never been to increase productivity. Productivity is seen as the means not the end. Only if the effect on productivity nullified the effect on well-being could these laws be regarded as failures. We should therefore try to look at their benefits and disadvantages to well-being now and in the future. This consideration will be highly speculative and involves a discussion of inherent dangers and safeguards as well as effect of productivity in a changing world.

The Workers Protection Act of 1949 and amendments of 1974 have done a lot of good in improving physical conditions of work. The success of the safety delegate depends on how sensible he is. So far comments have been favorable, and there has been no adverse criticism. An irresponsible safety delegate could result in a loss of productivity.

The Work Environment Act of 1978 will seek to improve psychosocial conditions for workers in the way they want. Much is expected of it. Is the nonspecificity of the act a weakness or a strength? Norway prefers to be specific. Then everyone knows where they are. A possible danger is that the working of the law is in the hands of bureaucrats with really no appeal to the judiciary. If the bureaucrats are biased in favor of one work situation, they may be unfair to other situations or employers. A reasonable bureaucracy and the changing times will determine the success of the act.

The 1977 Act of Codetermination at Work promised workers more say in running the factory or plant, which few of them wanted, but now risks disappointing them by giving the power to the trade unions instead. Modern management practices already include direct participation from many subordinates in decision making. The new law demands that decisions must be taken to trade union representatives and discussed all over again. Clearly the law could cause a great deal of loss of time and productivity. Further, since the law also applies to civil servants it allows the trade unions to influence decisions made by administrators appointed to execute the will of parliament. Theoretically, the law could be used to permit trade unions to influence a decision on the appointment of a chief editor of a daily paper or a judge on the basis that they are responsible for the staff working under them. Thus this act does not seem to be giving workers what they were led to expect, does give more power to trade unions, and does cause loss of time and productivity. The only safeguards are the responsibility of the trade unions and a monitoring or evaluating system that has been sponsored by the Workers' Environment Fund.

The Protection of Employment Act of 1974 has resulted in some curious situations where employees are hired for only 11 months and where unsuitable employees have not been given more suitable jobs and block good work in their present position. The redundancy provisions seem to be acting in a fair and acceptable manner.

The Act to Protect the Elderly and Handicapped appears to be working well. Sucess depends primarily on the responsible behavior of bureaucrats.

The Flexible Retirement Act is regarded as beneficial by all. It gives more independence and extra rights to individuals and does not impose extra duties. Costs will mount with inflation and the increasing proportion of old people in the population.

The effect of the law on equal opportunities as applied to government agencies has been reported after 6 months. In April 1977, the budget ministry gave a report on measures taken and planned by 247 government agencies in pursuance of the law. Most of the measures reported do not concern men and women other than those who are already workng in the specific agency concerned and most of the measures reported are hardly likely to have any effect on what is normally meant by equal opportunities in employment. The most common measures reported are discussions between the staff on sex roles, the principle of preferring internal promotion to external recruitment, taking turns making coffee, training typists to deal with administration, and having administrators type their own manuscripts.

The right of parents to decide who is to take leave during a baby's first year of life is not controversial. However some politicians are discontented that only about 5% of the fathers take such leave and want to put fathers under pressure to take the same amount of responsibility for their infant as mothers. This transformation of yesterday's right to tomorrows' duty is a common danger everywhere. It is particularly apparent in the equal opportunities area. Housewives are under cultural pressure to work. This inncreases competition—and disappointment—because there is already considerable unemployment. It causes frustration to elderly or middle-aged women who have devoted their lives to their families when they are now told that "one cannot develop one's personality unless one has a job." It causes discontent in typists when they are told that typing what somebody else has created is below human dignity. The possibility that new and untested methods of looking after children might be unsatisfactory for the children has not been adequately considered.

If applied in an irresponsible manner, most of the innovations discussed here are likely to reduce productivity and therefore profit margins. But the latter are needed to finance these innovative schemes.

The Volvo and Saab innovations on engine assembly have proved to be successful in terms of satisfying workers and employers and the quality of output and productivity. It should be remembered that these measures were proposed by all concerned, planned by all concerned, and carried out by all concerned, voluntarily.

The evaluation of the Kalmar plants is described in Pehr Gyllenhammar's text entitled *People at Work* and is based partly on interviews with over 100 blue- and white-collar workers at the factory plus the observations of a research team:

Nine out of ten workers participate in job rotation, and eight out of nine think it is a good way of working.

The workers are somewhat free to take breaks because of the buffers, but the original goals have not been met.

Decisions are delegated beyond the norm in traditional factories.

Nine out of ten workers want to take responsibility for the quality of the product, and feel that they are partly able to do so.

Final adjustments take more time than original plans anticipated.

The employees say the physical working environment is very good. Original goals for noise, light, and health service have been met, but they asked for certain changes regarding working postures and summer ventilation.

Productivity is as good as Torslanda, and the flexible layout will pay for the higher investment when it is used to full capacity.

Personnel turnover and absenteeism are around five percentage points lower than Torslanda.*

Nevertheless, there are some possibilities that this approach is not the best for all workers. There are indications that psychosomatic complaints have risen in some of those employed in the new way, and this is speculatively associated with unwanted increased demand "to make decisions."

Thus we can say that some of these innovations have undoubtedly increased well-being and some have not. The trend has been to give more power to the trade unions or to bureaucrats. The ordinary person, the ordinary worker, and the ordinary employer have, if anything, less power. There are inherent dangers in this and also in rights becoming duties.

Monitoring and evaluation—essential safeguards—are specifically proposed only for the law of Codetermination at Work. Profit

---

* P. G. Gyllenhammar, *People at Work* (Reading, Mass.: Addison-Wesley, 1977).

margins have recently diminished or become losses because of high costs, inflation, and falling external demands.

Sweden's unemployment at 1.6% is among the lowest in the world. While it is true that public employees comprise a large percentage of Sweden's employment force and therefore make cross-country analysis difficult, it is equally true that, even discounting that factor, Sweden has shown a remarkably low and stable unemployment rate. Some Swedish industry has dared to defy the basic tenets laid down by Adam Smith and indeed has shown success. The answers, as in all cases, are not clear, but some innovations have been tried, and it remains for other countries to venture forward.

PART TWO

INDUSTRY: POLICIES AND PRACTICES FOR INDUSTRIAL INNOVATION

7 The Environment for Industrial Innovation in the United States

Howard K. Nason

For the purposes of this discussion, innovation is the total process of creating, developing, and bringing to the market a new product or process. Usually, but not always, it is based on new technologies or new combinations of technologies.

That innovation is economically and socially desirable has been documented by a number of studies and generally is accepted. But because innovation involves technological change, it has become increasingly subject to pressure from members of society who have elected, for reasons of their own, to reject change as a socially desirable force. In his key study of the subject, Mansfield states, "Technological change is clearly an important factor in economic growth, both in the United States and in other countries, both now and in the past." Focusing on the research-and-development component of innovation he observed, "We may be underinvesting in particular types of R and D in the civilian sector of the economy, and the estimated marginal rates of return from certain types of civilian R and D seem very high."[1]

The importance to the national economies of innovative, new technology-based enterprises recently was documented by Arthur D. Little, Inc., in their report prepared for the Anglo-German Foundation for the Study of Industrial Society that analyzes experience in the United States, Great Britain, and Germany, including the effects of barriers and incentives.[2] To illustrate the long-term value of innovation, they cite the examples of three technological industries—television, jet aircraft, and digital computers. Nonexistent in 1945, they contributed more than $13 billion to the GNP of the United States and were responsible for the employment of 900,000 people just 20 years later. Hence the maintenance of a healthy proportion of innovation in the economy should be a matter of national concern, and research-and-development activity, since it is at the head end of the process, should be watched closely as an indicator of future trends. Recent observations afford little comfort.

Concern that innovation has diminished in recent years, at least in the United States, has stimulated much self-examination in both private and public sectors. In its February 16, 1976, issue the conservative magazine *Business Week* devoted almost eight pages to "The Breakdown of U. S. Innovation," expressing concern over the adverse economic effects of a decline in innovation and attemptng to examine some of the reasons for it.[3] The subtitle of the article, "No-risk, supercautious management is one of the prime villains," certainly is an oversimplification. The body of the text recognized that significant changes in economic and social environments have had adverse effects, some of them working through political actions whose side effects were not foreseen or adequately considered.

Concern that the technical base for innovation in the United States is eroding has been heightened by the publication of the National Science Board's "Science Indicators 1974."[4] Although the indicators do not completely agree, the general conclusion is that the "American performance has deteriorated in absolute terms" when compared with that of other innovation-oriented countries.

Also of concern is the finding that "industrial R and D is concentrated in a few industries and in a relatively small number of companies within those industries. Just 31 companies accounted for more than 60 percent of all R and D expenditures. Small firms those with fewer than 1000 employees) produced the greatest number of major innovations during the 1953–59 and 1960–66 periods, but large manufacturing companies (those with 10,000 or more employees) led in innovations in the 1967–73 period."[5] A favorable finding was that basic research contributes increasingly to technological innovation.

More currently, the trends survey conducted by the Industrial Research Institute's (IRI) Resarch-on-Research Committee indicates that industry funding of research is barely keeping up with inflation. Ninety-three companies indicated that their total expenditures for in-house research and development would be about $3.6 billion in 1977. The compounded average annual rate of increase in this total, from 1974 to 1977, was about 10%. The corresponding inflation rate was 8.3%. The change in number of research-and-development personnel from 1976 to 1977 was essentially zero. By the National Science Foundation definition, basic research amounted to $178 million in these firms, about 5% of the total. The growth over 1976 was only 3.9%, however, which is less than the rate of inflation for that period.

The situation with respect to government funding of research is even less encouraging. Data presented by Dr. Betsy Ancker-Johnson, former Assistant Secretary for Science and Technology, United States Department of Commerce, at the Industrial Research Institute's 1976 fall meeting, show a steady decline, as measured in 1967 constant dollars, in federal funding of basic research since 1968 and of applied research and development since 1967.[6] Industry funding of basic research, by the same measure, began a modest rate of decline in 1971. Its funding of applied research and development, after climbing steeply over two decades, more or less plateaued beginning in 1973.

The United States federal research-and-development expenditures will be less in 1978 than in 1968 based on constant (1972) dollars. This is shown in Figure 7-1.

Battelle Columbus Laboratories, in December 1976, forecast an increase of about 12.5% in the total United States outlay for research and development in 1977 over 1976, an increase of roughly $4.5 billion over $48 billion in 1976.[7] Of the 1977 estimated total, about 53% would come from the federal government, 43% from industry, and 4% from other sources. The actual work, however, would be performed approximately 14% by the federal government, 70% by industry, and 16% by other institutions.

Perhaps as important as how much is invested in research and development is how it is invested. Reliable data on how industry invests now are available in great detail as the result of accounting and reporting practices adopted by the Securities and Exchange Commission (SEC) and first mandatory for 1975. By examining 10-K statements filed with the SEC, a laborious task incidentally, a great deal of information can be obtained. *Business Week* devoted a major portion of its issue of June 28, 1976 to such an analysis and followed with an evaluation of the research-and-development management of General Motors Corporation, the largest corporate spender for research-and-development spending $6.5 billion.[8,9] Figures for 1976, with estimates for 1977, have been prepared by the Economics Department of McGraw-Hill Publishing Company.[10] They forecast an 11% increase in 1977 over 1976 to $30.6 billion.

72 Industry: Policies and Practices for Industrial Innovation

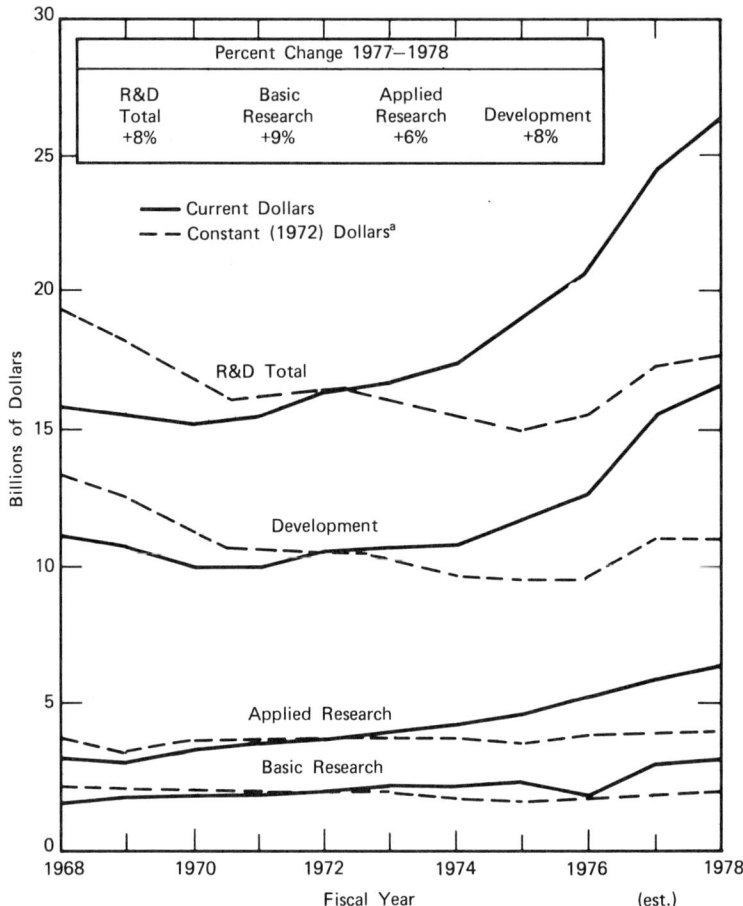

Figure 7–1. Federal R & D Obligations by character of work (fiscal years 1968–1978).

But the figures are not encouraging from the standpoint of innovation. Research and development aimed at innovation—new products, new processes, new plant investment—generally is down sharply. Major effort is being diverted into defensive research to protect existing investment, respond to imposed regulation, or to solve energy-related problems. The IRI trends survey referred to previously showed that the 93 firms responding expected to spend 11% of their total research-and-development effort in 1977 on energy-related research. Further, their research and development required solely in response to legislated regulations was growing at a compounded annual growth of 17.6%.[11] The McGraw-Hill study estimates that research and development related to environmental issues will grow at an annual rate of 38% from 1977 to 1980, with total spending going from $1.88 billion in 1977 to $2.6 billion in 1980.[12]

The automobile industry has been especially impacted by regulations that in many cases were conflicting. For example, regulations aimed at improved crash-survivability generally require increasing the weight of a vehicle, whereas those aimed at improved fuel economy can best be met by reducing weight. Also, reduction of emissions generally involves techniques that add weight and reduce fuel economy. By impacting individually and at different times, with little or no coordination between the agencies responsible, and little attention to the conflicts, the regulations produce great waste of effort and uncertainty.

Data from the automobile industry indicate that companies spend between 30 and 50% of their research-and-development expenses to meet mandated changes. General Motors found that in 1974 their total regulations-imposed cost was over $1.3 billion, involving the equivalent full-time effort of 25,300 employees.[13] The breakdown was as follows:

Area	Expenditures ($ millions)	Equivalent Full-time Employees
Regulation of Vehicles	884	17,500
Regulation of Plant Facilities	181	1,800
Reports and Administrative Costs	190	4,900
Occupational Safety and Health	79	1,100
	1,334	25,300

Such regulation-driven changes can have positive effects on innovation, though at great economic and social costs. For example, solutions of the weight-economy-emissions ambiguities have involved major design changes often using new or lighter materials.[14] The cost to the consumer is reflected in vehicle cost, which is expected to rise by 40% by 1983, assuming a 6% rate of inflation. Lewis and Gillette have commented on counterreacting aspects of regulation.[15] The viewpoint of the aerospace industry on the current state of innovation is reflected in a series of five coordinated papers that are recommended for study.[16]

The McGraw-Hill survey, referring to its projections for 1977, cautions that "a large part of the gain is centered in energy-related

R and D and not on new product research." For the chemicals sector alone it forecasts a 284% increase in energy-related work, from $44 million in 1976 to $169 million in 1977. Asked "How much of your 1980 sales will be in new products?" chemical firms responded that 19% would be. A year earlier they had predicted that 24% of 1979 sales would be in new products. About 59% of the industry effort for 1977 will be for improvement of existing products.[17]

In its 1976 annual report duPont comments that its research and development is "geared to achieving relatively short-term payout from product development and process improvement." About 66% is aimed at modifying existing products, and this is expected to rise to about 75% "in the years immediately ahead." In "a world economy increasingly characterized by high inflation and tight capital, further broad-scale diversification does not appear to be valid at this time," the company comments. Longer-range programs will "concentrate on products for which our present technical base gives us a competitive edge."[18]

The Battelle forecast also notes that a great deal of concern, has been expressed recently about the future implications of the shift in research-and-development from basic to more pragmatic research. The general nature of this concern has been that the reduction of research-and-development funding—and especially the reduction of support of basic research—is certain to reduce the longer-term viability of the U.S. economy.

The forecost comments that one of the most important reasons for the recent shift away from basic research is the growing need for defensive research and development. The nature of industrial research and development is changing because of growing governmental emphasis on environmental protection, the health and safety of the work force, and consumer safeguards. For several years now, more and more research and development effort is being directed directed toward these subjects—mainly in order to limit or reduce corporate liability under particular laws. Only a few years ago, most industrial research and development activity was directed to product or process improvements. Now it is directed to reducing or avoiding legal liability.[19]

Moreover, because of the uncertainties generated by recent inflation, the business community has altered its attitude toward many long-term investments, including research and development. Businesses are putting much more emphasis on short-term cost reductions than on long-term product and process improvements in business decisions relating to research and development.

The Environment for Industrial Innovation in the United States 75

The pressures are not confined to industry. Both government and academia feel increasing demands for relevancy, for identification of the payoff before work begins. The roots of this pressure go back to the "Mansfield Amendment," which prohibited the Department of Defense from supporting basic research not demonstrably related to defined missions and programs.

Comments made by leading research managers and industrialists at the 1976 fall meeting of the National Academy of Engineering in Washington reflect the climate of uncertainty and indirection currently enveloping the science and technology community. "The innovation process is in trouble in the United States. . . . Science and technology is less bold, less innovative, more timid. . . . Lack of teamwork between government and private enterprise is the No. 1 bottleneck in the nation's pursuit of research and development. . . . There has been a drastic decline in industrial support for research needed to provide the technological basis required to keep the United States competitive in world markets."[20]

To help meet the nation's problems and challenges, new ways must be developed for private industry and government to join together in "creative collaboration," Dr. William O. Barker, president of Bell Telephone Laboratories, told the meeting. Industry, working closely with universities, he said, must learn to "understand independently the macrosystems which we have too easily and casually assumed could be comprehended and assessed only by government.

"Dr. Simon Ramo, vice chairman of the board and chairman of the executive commitee of TRW Inc., criticized the trend of extreme insistence on a priori proven relevance for basic research and increasingly costly and cumbersome administrative and bureaucratic controls. Ramo said, 'Even the nation's most capable researchers spend much of their time defending their selection of projects and describing potential results, rather than in carrying on research. We are persisting in this tendency even with growing indications that the U.S. is losing its leadership position as to fraction of GNP being devoted to basic research.'

"In science and technology the government is the chief supplier of funds, biggest customer, sole regulator and rule maker, leading priority setter, and most influential judge of the results. But, for the government's billions of dollars of investment to yield a good return, a harmonious meshing must exist of government, private industry and university efforts. This is rare. Too often the government's approaches create handicaps, many of them very penalizing and most of them unnecessary."

The implications are obvious. We are sacrificing innovation to programs unrelated to growth, to investment, to increased employment, and to personal income, however otherwise desirable such programs may be. We are sacrificing programs aimed at long-term benefits to programs of expediency aimed at short-range payoffs. We might be able to tolerate such a situation if it is a temporary one, but the results of allowing it to continue for long would be disastrous to our world position, technologically, economically, and, eventually, politically.

The bicentennial report of the National Science Board identified the following concerns most often expressed by the principal sectors of the research commnuity.[21]

From the industry sector:

Government regulations and controls (unreasonable, not thought out, no cost/benefit/risk analysis). Absence of national science and technology policy, priorities or goals.

Near-term relevance is only research objective (due to government regulations or decentralization of research to profit centers).

General economic conditions, particularly inflation in salaries and laboratory costs, lead to decreases in fundamental research in industry.

Low public confidence in and/or poor image of science, technology, research or scientists.

Lack of availability of money, low profitability or obstacles to capital formation lead to decreases in fundamental research in industry.

Deteriorating patent protection or patent policy is a disincentive to industrial research and innovation.

Competing R&D functions (e.g., applied research or development in response to government regulations) decrease fundamental research in industry.

From the university sector:

There is pressure for applied research in preference to basic or pure research; projects are overly "targeted" or their subjects too minutely defined.

There is need for more continuity and stability in government funding of research; research grants should be longer.

More money in general is needed for research; there should be more basic research.

The public has a negative attitude toward science and technology.

Government (state, local, or federal) or one of its branches or agencies has a negative attitude toward science and technology.

More support for university research should be supplied at the institutional level.

A program of education or communication is needed to convince the public and government of the value of research.

There are excessive demands for accountability in the use of funds provided by government.

From the government sector:

Need for coordinated research policy at the national level involving long-range planning, commitments and priorities.

Increased emphasis on short-term research and neglect of basic research.

Overmanagement as evidenced by too many restrictions, especially on longer-term research.

Need for increased or stable funding.

Meeting public demand for justification of basic research programs with respect to mission.

Lack of Congressional or Executive support and understanding of basic research.

The three sectors are remarkably similar in their perception of key issues affecting research and development.

To this point our discussion has focused largely on the research and development-related aspects of innovation. But innovation involves much more, including engineering, plant design and construction, market introduction, and finance, all of which require investment for expenses, capital, human resources, natural resources, and energy resources. Every project competes for all of these resources, the total supply of which always is limited. Projects that consume these resources without adding substantially to the national output must be justified on the basis of different values than those traditionally employed for evaluating innovation and certainly not solely— or even mostly—on rational economic grounds. Since such values are extremely difficult to quantify, there is always danger that they will be overstated or emphasized disproportionately on emotional bases.

The Chairman of the duPont Company stated that duPont will have to spend $3 billion by 1985 to meet existing or proposed air, water, and noise laws and regulations, and that 75% of this expenditure "will buy no discernible improvement in the environment."[22] He suggests "periodic Zero Based Regulatory Reviews," to require only investment that actually are needed to achieve clear-cut additional public benefits.

Regulatory reform is an active issue before the United States Congress at the present time. The "sunset" approach (similar in principle, or in effect, to zero base review) has been proposed.[23] It is not at all certain, or even probable, that the proposed Regulatory Reform Act of 1977 will be enacted, or that it really would result in a more rational, systems-oriented approach to the regulatory process. Nevertheles, it deserves a try.

A good analysis of barriers to innovation appears in a study for the National Science Foundation made by the American Society of Association Executives.[24] The study categorized barriers into three distinct groups:

"Barriers related to regulatory policies and practices other than anti-trust and patent-related, and the present or future availability of raw materials and other resources.

"Barriers related to capital, anti-trust, corporate policies, human resources and design problems.

"Barriers related to markets and patents."

Highly important is their finding that "the mere phenomenon of *uncertainty* of future government policy was seen by more than 50% of the respondents as a deterrent." Pointing out that "no single policy, or set of policies, or no one particular initiative" would spur technological innovation "across the board," the study concludes with specific recommendations for action by *both* the public and the private sectors to achieve the sought-for objectives.

Bradbury puts the problem in perspective, emphasizing "the need to take all important facets of the task into account from the beginning of the innovations process; not only its technical viability ('it works'), but also its economic viability ('it pays'), its sociological viability ('it is socially acceptable'), and its desirability ('cost–benefit analysis').—To be pursued effectively, there is need for a balanced attack on the centers of uncertainty."[25]

"The very nature of innovation is uncertainty. But the economists' studies which show that, even allowing for failure, innovation has both private and social returns, should give heart to those who seek more efficient and effective ways to meet man's—and the U.S. economy's—changing needs."[26]

Innovators understand most forms of uncertainty. Technical uncertainty, market uncertainty, economic uncertainty—these are believed to be manageable. But political uncertainty—the intervention of the government through regulation—is generally regarded as unmanageable.

8 New Technology-Based Firms

Arthur D. Little, Ltd.

Objectives

Arthur D. Little (ADL) were asked by the Anglo-German Foundation for the Study of Industrial Society to make a comparative assessment of the environment for new technology-based firms in the United Kingdom and the Federal Republic of Germany. The aims of this study are to provide a detailed analysis of the environmental factors in each country that influence the development of new technology-based firms, and to make recommendations on how the creation and growth of such firms might be encouraged. In providing this insight into the practices of industrial society and government in the two countries and in recommending ways of promoting technological innovation, this study contributes to the objectives of the Anglo-German Foundation, which are:

- To promote the study and to deepen the understanding of modern industrial society and to advance the knowledge of the British and German peoples about that society and about ways and means of resolving the problems that arise in it.
- To advance and foster education and knowledge in the two countries in the fields of science, technology, commerce, economics, sociology, and the arts with a view to promoting and stimulating development of industrial society in a manner most beneficial to the community.

From "New Technology-Based Firms in the United Kingdom and the Federal Republic of Germany." A Report prepared for the Anglo-German Foundation for the Study of Industrial Society by Arthur D. Little, Ltd., June, 1976.

Scope

Before starting out on the research program, it was essenital to establish a working definition of a new technology-based firm (which will be referred to by the initials NTBF throughout this report). Our analysis begins somewhat downstream in the process of innovation at a point where the invention has already been made and an individual has decided to start up a new business based on the technology. At this point, the inventor/entrepreneur finds himself in the business environment where the issues confronting him are no longer "how to invent" but questions of incorporation, finance, taxation, patents, production, marketing, and competition.

For the purposes of this analysis, an NTBF is defined by the following characteristics:

- The firm must be new (i.e., it must have been set up since January 1, 1950).
- The firm must be technology based. This part of the definition is inevitably the most difficult because there can be no precise definition of technology. There will always be a gray area of businesses that some people will call technological and other people will not. For the purposes of this study, we shall classify a firm as technology based if the business is based on a patented invention or the business has substantial technological risks in addition to the normal commercial risks. An NTBF, therefore, has a much higher risk of failure than an ordinary business.
- The firm must have been set up by an individual or group of individuals. The person who starts up the business will be called the inventor/entrepreneur. We exclude from our definition, and from this study, all technology-based firms set up as subsidiaries of established companies.
- The firm must have been set up for the purpose of exploiting an invention or technological innovation. We, therefore, exclude from our analysis the multitude of nontechnological firms that happen to make innovations in the course of their ordinary business.

It should be clear from this definition that we do not study small firms in general and that we do not study technology-based firms in general. We do study new firms that have been set up by individual inventors/entrepreneurs for the specific purpose of exploiting an invention or technological innovation.

General Approach*

To provide a framework for the study, we considered the major factors in the environments of the two countries that, directly or indirectly, affect the creation and growth of NTBF's. These factors are illustrated in Figure 8-1. This report considers four of the six factors: the national research-and-development system, the role of venture capital, taxation and forms of incorporation, and the patent system. The two remaning factors—economic conditions and cultural attitudes—are very general influences. It is not a practical proposition to analyze these two enormous subjects and directly determine their

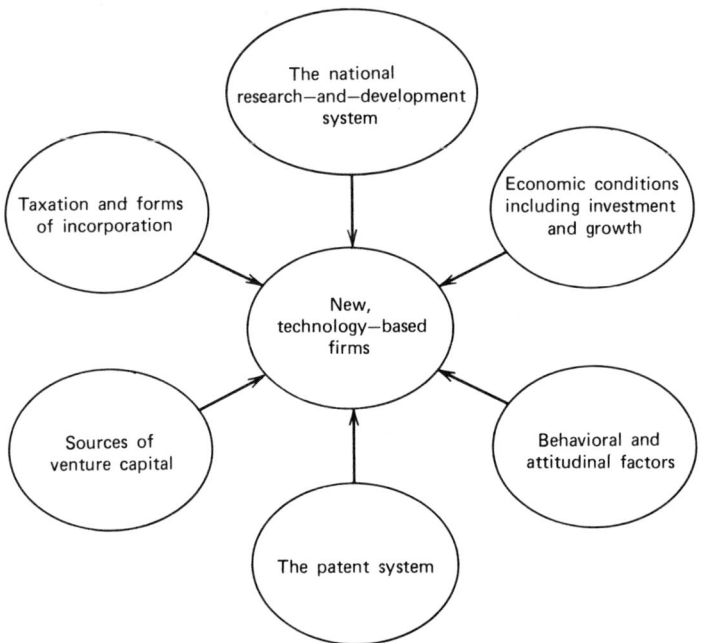

Figure 8-1. Environmental factors affecting the creation and growth of new, technology-based firms in the United Kingdom and the Federal Republic of Germany.

---

*For detailed approach see, "New Technology-Based Firms in the United Kingdom and the Federal Republic of Germany," a Report Prepared for the Anglo-German Foundation for the Study of Industrial Society (Arthur D. Little, Ltd., London, June 1976).

impact on NTBF's. The effect of economic conditions is referred to later. Cultural and attitudinal factors are discussed in our analysis of the national research-and-development system.

The method of research included a review of the published literature, about 100 interviews with experts in the various subject areas, consultations with tax and patent experts, and extensive field research with NTBF's in the United Kingdom and the Federal Republic of Germany.

Conclusions

The number of NTBF's in both the United Kingdom and the Federal Republic of Germany is low, and their performance has been unimpressive, particularly in comparison with the United States. With a few exceptions, such as Racal Electronics, Ltd. and Nixdorf Computer AG, NTBF's in the two countries have demonstrated no particular success whether measured in terms of numbers, size, growth, or contribution to GNP or employment.

The number of NTBF's set up since 1950 and still in existence is about 200 in the U.K. and probably fewer in the Federal Republic of Germany. The total sales of these firms were over 200 million in each country in 1975. By contrast, there are several thousand NTBF's in the United States and their sales run into billions of dollars.

The low level of NTBF activity in the United Kingdom and the Federal Republic of Germany has significant long-term implications for both countries. They are neglecting:

- An important channel for the exploitation of technological innovation
- The development of a new generation of modern industries that are needed to provide future employment and exports
- The value of NTBF's in maintaining a competitive environment in the face of the increasing power of major corporations.

The relatively poor showing of NTBF's in the United Kingdom and the Federal Republic of Germany is not surprising in view of the fact that the environmental factors are considerably less favorable than in the United States. Favorable factors in the United States include:

- A very large domestic market conducive to rapid growth and development.
- The availability of private wealth as a source of seed capital for the start-up of new ventures
- A fiscal framework which encourages the flow of private risk capital into new ventures
- The existence of an active market for the trading of shares in new ventures (i.e., the over-the-counter market)
- A prevailing attitude in society-at-large, which encourages entrepreneurship
- Greater mobility of individuals between academic institutions and private industry
- The behavioral and attitudinal character of American scientists, many of whom are willing to set up their own businesses in order to exploit their technical knowledge
- A large and active government expenditure program in high-technology areas that provides significant opportunities for NTBF endeavor, particularly through government procurement programs

The low level of investment and economic growth in the U..K has had an adverse effect on the creation and growth of NTBF's. Yet the excellent economic performance of the Federal Republic has not led to the creation of large numbers of successful NTBF's in that country. We conclude that although bad economic conditions can have a negative effect on the number and performance of NTBF's, a favorable economic climate is not alone sufficient to generate these firms.

The pattern of factors affecting NTBF's is different in the two countries. In the Federal Republic, the tax system, the patent system, and economic conditions are generally more favorable to NTBF's than in the U.K. Yet, until the creation of Deutsche Wagnisfinanzierungs GmbH in 1975, there were no venture capital institutions in the Federal Republic and no equivalents to Technical Development Caiptal Ltd. (TDC) and the National Research Development Corporation (NRDC), which have financed a significant number of NTBF's in the U.K. TDC and NRDC have been positive factors in the British environment, but their impact has been more than outweighed by unfavorable economic conditions and the disincentive effects of taxation.

Three negative factors are common to both countries. Cultural and attitudinal factors among academicians, government scientists, and research institutions are unfavorable toward technological entrepreneurship. In the United Kingdom, government research-and-development expenditure has consistently neglected NTBF's, and, until recently, the same has been true in the Federal Republic of Germany. The growth of NTBF's in both countries has been restricted by the fragmentation of the European market.

In the United Kingdom and the Federal Republic, only a minute fraction of government research-and-development exependiture is available to, or actually received by, NTBF's. This issue has attracted virtually no attention in the United Kingdom, whereas in the Federal Republic of Germany there has been substantial criticism of the fact that small firms do not receive a representative share of public research-and-development funds. As a result, the German government is very much aware of this issue, whereas the British government is not.

In both countries there is a considerable amount of official prejudice against small firms in public research and development. Both governments prefer to achieve their national research-and-development objectives through large established companies.

In both countries government research establishments are poor incubator environments for NTBF's. Private industry is probably the most fertile source of inventors/entrepreneurs, with the universities second. The main reasons for the lack of spin-off of inventors from government establishments and universities are: cultural and attitudinal factors on the part of scientists and their institutions and the fact that most of their research is of such a type and scale that it cannot easily be exploited by setting up a small firm.

The British government led the way in developing a policy for the exploitation of inventions by setting up NRDC as early as 1949, but in recent years the government has given virtually no attention to the process of innovation. By contrast, in the Federal Republic innovation policy has been the subject of increasing interest in recent years. As a result, the Program for Sponsorship of Innovation has been established; the Key Technologies Program is orienting more of its funds toward small firms; the Deutesche Wagnisfinanzierungs GmbH has been created; and a large number of research studies have been carried out.

In the Federal Republic the different types of support that assist the creation and growth of NTBF's are divided between six organizations and programs: ISI-ARPAT, Garching Instrument, Patentstelle für die Deutsche Forschung, Program for Sponsorship of Innovation, European Recovery Program, and Deutsche Wagnisfinanzierungs GmbH. It is doubtful whether these organizations can achieve their full impact while their efforts remain so fragmented.

In theory the NRDC in the U.K. is an ideal public institution for sponsoring NTBF's because the various types of assistance are combined in one place, it is a large organization with substantial technical knowledge, it has the ability to handle patents, and it has large sums of money available. However, it is doubtful whether NRCD is really interested in providing venture capital for NTBF situations.

Equity investments in NTBF's are a very small part of NRDC's activities. The number of occasions where NRDC is willing to put equity into new small firms has been declining and there were none in the first half of 1976. The National Research Development Corporation's equity investments in NTBF's have not been a success so far, and the corporation is not attracting good proposals from inventors/entrepreneurs.

Although the Deutsche Wagnisfinanzierungs GmbH is very different from NRDC in its purpose and scope, it will probably encounter some of the same problems as NRDC and should learn from its experience.

Venture capital for the financing of NTBF's is more easily available in the United Kingdom than in the Federal Republic of Germany. There are more than a dozen institutions in the United Kingdom that provide venture and develeopment capital for small and medium-sized firms, though NRDC and TDC are the only British institutions that really focus on NTBF's. In general, traditional sources of finance in the U.K. are receptive and positive toward new and developing ventures with high-growth potential. However, the impossibility of floating new issues on the Stock Exchange during the past 3 years has removed one of the major incentives for venture capital investment.

In the Federal Republic of Germany, no specialized institutions exist to provide venture capital, except the Deutsche Wagnisfinanzierungs GmbH (DWFG), which was set up in 1975. Traditional sources of finance are unreceptive to the needs of new firms for unsecured financing with the exception of short-term overdrafts. The impossibility of going public removes the incentive and the opportunity for equity-oriented venture capital.

The DWFG is still in its early stages of operation and is just starting to implement its investment strategy. The initial capital of DM 10 million is probably too small a base on which to run a national venture capital operation. The DWFG needs to grow rapidly in order to reach a viable critical mass. It is also very important that the DWFG should be dominated by managers with business, industrial, and technical backgrounds and not by the banker mentality.

There are three principal venture capital institutions operating in the United Kingdom and the Federal Republic of Germany that focus on technology-oriented investments: Technical Development Capital Ltd., European Enterprises Development SA, and Scienta SA. Although all three are continuing in business, none has been a great success so far. Two of the main problems confronting venture capital operations in the U.K. and the Federal Republic of Germany, and the rest of Europe as well, are the lack of an adequate secondary market similar to the over-the-counter market in the United States in order to realize capital gain and the difficulty NTBF's have in achieving rapid growth and a means of realizing capital gains, venture capital cannot succeed.

The start-up of new technology-based firms is highly dependent on the existence of private wealth to provide the initial seed capital. Fiscal legislation is, therefore, very important because it determines the flow of private wealth into new firms, including NTBF's. German experience with tax shelter vehicles shows that tax incentives can have the effect of channeling enormous sums of private savings into various types of investment.

Very high rates of personal taxation in the U.K. have two adverse effects on private investment in new businesses. On the one hand, they make it extremely difficult to accumulate private savings. Then they act as a disincentive to investment of those savings in high-risk/high-return ventures. A private individual considering investment in an NTBF will look at it on the basis that if it fails he will lose all his money, but if it is really successful his marginal tax rate on dividends from the business may be 98%. A scientist may well come to the conclusion that the rewards after tax are too small to justify risking his personal assets, his career and his pension. Capital transfer tax and tax laws that take the value out of stock options are further major disincentives not studied in this report.

The situation in the Federal Republic is more favorable. Income tax rates are much lower. There is no surcharge on investment income. The GmbH & Co. KG enables investors to be taxed as partners while benefiting from limited liability. Losses made by a GmbH & Co. KG can be used as a tax shelter, that is, the losses of the business can be charged against tax on the investors' income from other sources.

In the area of coporate taxation, the Federal Republic of Germany is less favorable than the U.K. In the Federal Republic of Germany, retained earnings are penalized by a higher rate of corporation tax. The total amount of coporate taxes paid by a GmbH that retains all its income can be about 62% of profits compared with 52% in the U.K. This system disfavors NTBF's because they tend to retain most of their earnings.

The British system of capital allowances is of great benefit to the cash flow of an NTBF. In effect a British NTBF need pay no corporation tax as long as its capital expenditure (on most types of assets) is greater than its pretax profit. By contrast, the combination of lower depreciation allowances and a 62% tax burden on retained profits have a very adverse effect on the cash flow of a GmbH. As a result, the German company will have greater cash flow problems, a higher level of debt, a higher level of interest charges, and a lower level of earnings than its British counterapart.

It is paradoxical that the British system of coporation tax greatly benefits the cash flow of the NTBF itself, whereas high rates of income tax are a serious disincentive to individuals who might set them up. The converse is true in the Federal Republic of Germany where the disincentive effect of personal taxation is much lss, but the tax system has a very damaging effect on an NTBF's cash flow.

Although in both countries the tax authorities have made a token attempt to relieve the burden of corporation tax on small companies, these special provisions are not effective because the limits on the size of the company are too low.

Many NTBF's are not based on patented inventions and even in those firms which are, the technical knowhow of the inventors/entrepreneurs is often of much greater value than the strength of his patents, because the success of the NTBF depends on his ability to develop second and third generations of products rapidly in order to grow and remain competitive.

Patents can have both positive and negative effects on the creation and growth of NTBF's. On the one hand, they hamper the spin-off process because employers, both public and private, are entitled to claim the right to inventions made by their employees. On the other hand, patent protection gives security to the independent inventor and helps NTBF's to attract outside capital. Patents can also be of strategic value to an NTBF in a number of other ways.

However, the patent system is a specialized branch of the law, requiring a level of expertise that is often beyond the reach of an

NTBF. What is really lacking in the U.K. and the Federal Republic of Germany, as well as other countries, is the educational opportunity for managers in general to learn the real value of patents and the use of licensing in corporate development.

There are significant differences between the British and German patent system, but most of these differences have little effect on NTBF's. The one difference that is most relevant to NTBF's is the German Law of Employee Inventions. Under this law an employee inventor can in effect claim the right to exploit his invention if his employer does not exploit it. This part of the law creates ideal spin-off conditions. If an employee is frustrated because his company refuses or fails to exploit his invention, he can take it away and set up his own firm to exploit it. By giving this right to employees, the law also puts pressure on big companies to exploit all their inventions; otherwise they may lose them.

In the United Kingdom, unused patents are available for licence fairly freely, and there are provisions in the law to enforce compulsory licenses under unused patents. However, this is not the same as giving employee inventors a prima facie and specific right to claim inventions that their employers have refused or failed to exploit. If the German practice were adopted in the U.K., it would probably be used in no more than a handful of instances each year. But it would open the door to some employee inventors who want to exploit their inventions by setting up their own businesses.

Recommendations

There has been little concerted effort in th United Kingdom or the Federal Republic to create a favorable climate for the establishment and development of NTBF's. The few initiatives that have been taken in this direction have been confined to government sponsored direct measures such as the Program for Sponsorship of Innovation in the Federal Republic, and financial assistance from the NRCD in the U.K. No effort has been made to use tax incentives to encourage the creation and growth of NTBF's.

We believe that there are significant opportunities in both countries to stimulate the creation of NTBF's and improve the environment for their development, although certain underlying constraints such as cultural attitudes and the fragmentation of the European market will not change in the short term.

There are basically two different types of assistance that can encourage the creation and growth of NTBF's: direct measures and indirect measures.

Direct measures include specific programs designed to benefit specific types of firm; these programs usually require the expenditure of government money. Examples include:

- Grants or loans to companies that qualify because of their size or because of their involvement in research and development
- Finance provided by the NRDC in the U.K. and the DWFG in the Federal Republic
- Assistance with patents provided by ISI–ARPAT and the Patentstelle für die Deutsche Forschung in the Federal Republic

Indirect measures, which can also be referred to as nontargeted policies, consist of providing a favorable environment in which the inventor/entrepreneur has the incentive, the opportunity, and the resources to start his own business. This favorable environment includes:

- A high rate of economic growth and industrial investment
- Cultural attitudes that encourage individuals to make money out of their own businesses
- Behavioral factors that motivate the individual to start up on his own
- Availability of finance from private sources, financial institutions and the capital markets
- Tax incentives, or the absence of tax disincentives

One of the main conclusions from this study is that indirect factors are much more effective in encouraging the creation and growth of NTBF's than direct measures. Direct measures are normally a poor substitute for a favorable indirect environment. Although in the recommendations that follow we suggest some improvements in the government's direct measures, we feel very strongly that indirect measures particularly in the field of taxation, are inherently more effective in encouraging technological entrepreneurship.

As a general principle, the governments of the United Kingdom and the Federal Republic of Germany should do more to channel their national research-and-development efforts into NTBF's. A higher proportion of government research and development expenditure should be allocated to small firms (i.e., those employing less than 200 people).

The British government should follow the German example and regularly collect and publish figures to show government research-and-development expenditure in private industry by size of recipient firm.

In the United Kingdom and the Federal Republic of Germany, publicly funded research establishments should be encouraged to cooperate on projects with small firms; the establishments should be enabled to invest a small proportion of their budgets in NTBF activities associated with their fields of research.

In the Federal Republic, the Ministry for Research and Technology should aim more publicity toward small and medium-sized firms to inform them about the various forms of public sponsorship for technological innovation. The application procedures for these forms of public sponsorship and the formalities for controling the funds should be simplified for small firms.

The Department of Industry Research-and-Development Requirements Board in the U.K. are at present willing to consider applications for research and development funds from NTBF's. The Boards should now publicize their interest in receiving such applications and assure that they are organized to provide an effective and timely response.

The NRDC should establish a permanent venture capital department whose primary mission should be to finance NTBF's. This department should carry out a positive search for NTBF investments and should not hesitate to provide funds to finance the working capital and production facilities required by NTBF's.

There is a strong case for making the small Firms Division of the Department of Industry into an independent government agency with the power and the responsibility to comment publicly on the effect of government policy, legislation, and taxation, on small firms.

One of the most effective ways to encourage the creation and growth of new businesses in the United Kingdom is to reduce the disincentive effect of high rates of personal taxation. We, therefore, recommend a major reduction in the rates of income tax on earned income and investment income, bringing these rates closer to those operating in the Federal Republic.

The British and German governments should adopt tax provisions similar to Subchapter S and Section 1244 in the United States. In other words, small limited companies in the U.K. and small GmbH's in the Federal Republic of Germany should have the option to be taxed as partnerships. Shareholders in these companies should be permitted to deduct capital losses on their shareholdings from their personal income before tax is assessed. These provisions should be restricted to companies engaged in manufacturing.

In the Federal Republic of Germany, small firms engaged in manufacturing should be given 100% first-year capital allowances on their purchases of plant and equipment, similar to those operating in the U..K

In the United Kingdom, the small companies rate of corporation tax should be extended to those companies whose pretax profits do not exceed £100,000 with tapering relief up to £150,000. This additional relief should be restricted to companies engaged in manufacturing.

In the Federal Republic of Germany, there should be a substantial reduction in corporation tax on the retained profits of all companies engaged in manufacturing whose pretax profits do not exceed DM 500,000 with tapering relief up to DM 750,000.

We recommend that, following the provisions of the Law of Employee Inventions in the Federal Republic, employee inventors in the United Kingdom should be given the right, with respect to inventions they have made in the course of their employment to exploit patents their employers refuse or fail to exploit.

In order to encourage inventors to exploit their own inventions, we support the intention of the British government to invalidate clauses in contracts of employment that attempt to give employers the right to patent inventions made by an employee outside the course of his employment.

9 Innovation in Medium and Small Industrial Firms

Karl A. Stroetmann

In recent years, most industrialized countries have been experiencing economic recessions, high unemployment rates, and changes in international trade flows to a degree unaccustomed to since World War II. At the same time, new social demands have arisen to further strain available resources and require new solutions to satisfy them. In the past, innovations have been a major factor shaping economic growth. Innovations, particularly technical innovations, are also widely regarded as a major means for overcoming present problems and for regaining economic growth rates sufficient to satisfy ever-increasing private wants and social needs.[1] When discussing technical innovations, this term should be understood to include the processes from conception or generation of an idea to its wide-scale utilization by society including activities involved in the creation, research, development, and diffusion of new and improved products, processes, and services for private and public use.[2]

It is widely believed that smaller firms are particularly hard hit by worldwide changes in the international division of labor and related structural change. Though the percentage of industrial firms having less than 50 employees decreased slightly in West Germany between 1965 and 1973, and the relative percentage of people employed in companies having fewer than 200 employees also declined somewhat, no general tendency can be identified.[3] Rather, the development differs considerably from one branch to another. For example, in leather production, imports and substitute materials led to a loss of more than two-thirds of all jobs from 1963 to 1973. At the same time, companies having fewer than 200 employees increased the relative percentage of working places provided by about 70% (from 50% to more than 85%), whereas out of six firms having more than 500 employees only one—a smaller one—survived.[4] Smaller firms, by rapidly and flexibly adjusting to changing market conditions and new demands through offering innovative products and services cannot only survive but even prosper—if they are willing to and have the capability to innovate.

The German federal government has more and more recognized the particular role small and medium-sized companies can play in developing new products, providing new jobs, and securing economic growth. In the 1960s, promotion of industrial research and development was mostly concentrated in areas such as nuclear energy, space technology, and data processing—fields that largely lend themselves only to activities by large enterprises. Since the early 1970s, this has changed slowly. "The Federal Ministry for Research and Development endeavors unflaggingly to assist small and medium-sized firms with projects serving to better utilize and further extend specific experience. Those promotion programs which were initiated in 1970 in the field of technological research and development (i.e., in optics, metrology, materials development, and health technology) are a good starting point for such assistance."[5] Other measures followed, and a comprehensive innovation policy concept to support small and medium-sized enterprises is due to be presented by the end of 1977.

The Changing Environment

Entrepreneurs and managers, whether in small or large firms, are faced with a fast-changing environment. This change threatens the survival of many companies but, at the same time, opens up new opportunities. Technical progress is a major factor of this change as new techniques are developed and applied, known techniques diffuse into new areas of application, new combinations of generally available techniques lead to the creation of innovations, and old techniques are replaced by new ones.

For example, the development of the microprocessor (MPU) has already led to a small revolution in several industries. New or improved products incorporating them are now on the market such as smart video games, electronic watches, smart scales, and so on.

Hierarchical computer systems to control complicated production processes became feasible and for small business new markets opened up: "Development time is so short for a smart product now and the entry costs are so low that there will be myriad examples of new companies spawning, with bright, young fellows developing MPU-based products."[6] At the same time, many small companies producing mechanical components and devices are threatened unless they can adapt to this new technology or develop new products for different markets.

Worldwide economic change is another important factor affecting many small firms. The international division of labor and the comparative advantage of countries are changing due to changing exchange rates, increased raw material prices, changes in industry location advantages (often related to environmental concerns), and increased industrialization of developing countries. Demand for many standard consumer products seems to reach a certain level of saturation, and in the production of investment goods also a trend toward more specialized or flexible machinery can be observed.

These developments again mean new opportunities for small and medium-sized firms. Their change lies in adapting to these new trends in demand, in opening up new markets for high-quality specialty products, in making use of new materials to meet new needs of industry and consumers, and in offering complete systems for specific demands rather than producing standard products larger firms can manufacture more efficiently. To be able to achieve this, firms will increasingly have to make use of the specific comparative advantages of highly industrialized countries such as the availability of a highly trained and experienced labor force; an abundant reservoir of technical know-how; a developed research, development, and innovation capacity; and a sophisticated scientific–technological infrastructure.

Finally, the impact of social change on smaller enterprises should not be overlooked. The installation of new pollution control equipment, the improvement of working conditions, the increasing financial burden for social security payments, or the adjustment to new standards and regulations often pose more difficulties to smaller firms than to large ones.[7] But again, new demands open up new markets, and smaller firms can play an important role in inventing better pollution control and monitoring devices, in developing safer production equipment, or in organizing new support services in the health field.

Some Particular Problem Areas

Much has been written about the innovativeness of small and medium-sized firms vis-à-vis that of large ones.[8] I do not want to continue this discussion. Suffice it to say that both types of firms play their own role in the process of technical change, often depend on each other for and exchange technical know-how, and, through the dynamics of technical and economic change, vary in (relative) size over time, grow, decline, and even die. If for political reasons, governments decide to stimulate innovation in smaller enterprises, the identification of particular problem areas in need of and susceptible to public support becomes the central issue, not whether smaller firms are the major source of technical progress. Let me, therefore, briefly pinpoint some weak areas research and empirical evidence allows us to ascertain before outlining some implications for the promotion of technical innovations in small and medium-sized business.

Management. Innovation problems of smaller firms are usually related to or are an outgrowth of more general problem areas.[9] Poor innovation management is such a field. First of all, top management has to be motivated to innovate, but the innovation propensity is rather limited in many smaller enterprises. Firms led by an autocratic, conservative owner will find it difficult—if not impossible—to react adequately to a fast changing environment. In addition, innovative efforts should not be undertaken on an ad hoc basis but need to be planned well ahead and to be integrated into overall corporate goals. To reduce the risk of being surprised by changes in the technical, economic, or social environment, the development of a corporate strategy and of longer-term plans to implement and execute such a strategy is, therefore, mandatory. Again, this is a particularly weak point in almost every small enterprise.

Manpower. Undertaking research, development, and innovation requires qualified scientists and engineers. In many instances a team of experts possessing technical know-how and experience in diverse fields is needed to develop new or improved products and processes. The proliferation of electronic components into thousands of different applications persuasively illustrates this. Small and medium-sized firms, which usually do not have formal research-and-development departments and which can afford to spend only small amounts (in absolute terms) on research and development or design and development, have considerable difficulties in attracting and financing, on a permanent basis, one or several qualified scientists and engineers. As the data in Table 9-1 indicate, both the absolute number and the relative percentage of professionals in research and development employed by enterprises having less than 500 employess declined considerably over the 10-year period from 1964 to 1973 in Germany.

Information. For a rational planning and executing of innovational efforts, information on many different aspects are needed: technological developments, sources of technical assistance, economic developments, market situation, government policies, and promotional measures, to mention just the most important ones.

Table 9-1. Professionals in Research and Development in German Enterprises by Size of Firm, 1964–1973

Size of firm (employees)	Year	Professionals	
		Number	Percent of total
Less than 500	1964	576	5.2
	1965	443	3.7
	1967	425	3.4
	1969	289	1.7
	1971	394	2.1
	1973	390	2.1
500–1,999	1964	1,002	9.1
	1965	955	7.9
	1967	1,254	10.0
	1969	1,169	7.1
	1971	1,842	9.9
	1973	1,728	9.3
2,000 and more	1964	9,472	85.7
	1965	10,643	88.4
	1967	10,827	86.6
	1969	15,127	91.2
	1971	16,443	88.0
	1973	16,397	88.6

Source: H. Echterhoff-Severitt et al., Forschung und Entwicklung in der Wirtschaft (Essen: Stifterverband für die Deutsche Wissenschaft, 1977), p. 22.

Because of their size (as well as other factors), smaller firms are at a disadvantage in collecting and analyzing information. Research indicates that they are mostly not even aware of the serious information gaps they have to cope with,[10] and there is a great need to improve existing and to establish new sources to meet small firms' information requirements.

Finance. To innovate is an often costly and usually risky undertaking. Research itself is not always that expensive, but development, preparation of prototypes, testing, production start-up, and market development often require amounts beyond the means of smaller firms. And when they innovate, they normally cannot diversify the risk of this activity by embarking on several subjects at the same time. Both financial constraints and risk lead smaller firms to engage in less-innovative and therefore less-expensive, short-term projects as the data in Table 9-2 suggest. The fact that smaller firms show a particularly high rate of fluctuation with respect to research-and-development expenditure may also be partly attributed to these problems.[11]

Research and Development. The research-and-development expenditure data collected by the German Stifterverband indicate that the research-and-development participation rate of smaller firms—with the exception of those in mechanical engineering and possibly also in precision instruments and optical goods—is very low in Germany.[12] Because of the difficulties already mentioned and the resources required to entertain a minimum research-and-development capacity in many fields of technology, this is not surprising. Instead of performing research and development in-house, cooperative and/or contract research and development are often proposed as a solution to the difficulties of smaller enterprises to sustain a permanent research-and-development capacity. However, presently they spend very little on external research and development.[13] The reason for this state of affair is not difficult to identify. Empirical research showed that small and medium-sized enterprises do not place research contracts to any significant extent not due to limited financial resources but rather due to their limited technical and organizational capacities.[14] They are often unable to formulate their technical problems so that they can place a formal research-and-development contract, the results of which can be usefully transferred back into the company. Equally important, staff to coordinate and supervise such contracts and to spend sufficient time with the

Table 9-2. Annual Research and Development Expenditure and Average Duration of Innovation Process by Size of Firm[a]

Firm Size	Annual Turnover (million DM)					
	<0.5	0.5–0.9	1.0–4.9	5.0–19.9	20.0–49.9	50.0 and more	Average
Average Annual Research-and-Development Expenditure (thousand DM)	14.3	29.9	55.4	183.6	363.3	891.1	164.0
Average Duration of Innovation Process (mo)	10.4	11.1	13.8	19.3	21.3	18.0	15.6

Source: Industrie- und Handelskammer Koblenz, *Kein Technischer Fortschritt ohne Mittelstand* (Koblenz: November 1975), pp. 9–10.

[a] The data are from a survey of 751 industrial firms having 10 or more employees of which 377 returned completed questionnaires. Of these firms, 155 undertook some kind of innovation efforts.

institution performing the work is frequently missing. In addition, adapting research-and-development results and transforming them into marketable products requires usually an internal innovation capacity that is often lacking.

Similar problems have surfaced with respect to research and development performed by industrial research associations. Just supplying information on the results of certain research-and-development projects puts smaller firms at a clear disadvantage vis-à-vis larger enterprises that have sufficient in-house capabilities to digest, adapt, and further improve on such work. To assist smaller firms in this respect, the establishment of special service divisions in industrial research institutes has proved helpful; in addition, the research team should be held responsible for the transfer of the technology developed by it.[15]

Licensing is another way to obtain research-and-development results. Smaller companies play both as licencees and as licensors, a minor role when compared to large firms. Because of their usually limited research, development, and innovation capabilities they are neither as offerers nor as takers of new technology very interesting partners.

Promoting Technical Innovations in Small Business

What are the implications for government policies intended to stimulate and support innovation activities in small and medium-sized industrial firms? With respect to many factors playing an important role for a firm's ability to innovate, small and medium-sized companies are at a clear disadvantage vis-à-vis large corporations. But supporting smaller firms can be justified on several grounds:

Technical: In many areas of technology and in many industrial sectors small firms have significantly contributed to technical change. There is no reason to believe that this will change significantly if the peculiar disadvantages of smaller firms in the innovation process can be alleviated.

Economic: Smaller firms are an important component in competitive markets, which in turn seems to stimulate innovation.

Social: In many instances, new social demands burden smaller firms more than large enterprises. Particular support to allow the introduction of, for example, safer machinery, may therefore be necessary.

Political: Small business represents an important political force that governments cannot afford to neglect.

No easy generalizations about innovation processes in smaller firms can be made. Though several weak points can be identified, their factual importance as a hindrance to innovation varies considerably from industry to industry and for firms of different sizes. An individual firm's competitive position, the impact of international economic change on its markets, and other factors also have to be considered. As a consequence, if governments want to support smaller firms, only a host of general measures designed to attack different problem areas simultaneously or programs geared specifically to the weak areas of a particular industry or subsection of it will, in all probability, be successful. One may even argue that a combination of some general with industry-specific measures should be most efficient.

Management. In Germany, education in the management of innovation is virtually missing both at the universities and through continuing education institutions. Many innovation problems of smaller firms are at least partly related to weak management, and money spent on improving the management of innovation may show a higher payoff in the longer run than most other government programs. If managers know how to better plan, organize, and execute innovation programs and how to market the innovation, government efforts to subsidize research and development through various means should show a considerably higher multiplier effect and wider impact than up to now. The innovation center experiment of the American National Science Foundation and the planned application of this concept by the U.S. Small Business Administration should be closley watched in this context.[16]

Manpower. This is a particularly weak area. Since expenditures for research-and-development personal are the main component of total research-and-development expenditure, subsidizing total research-and-development expenditure in smaller firms should be a means to assist them in this problem area. A special subsidy on new research-and-development employees may prove even more helpful.

Information. Information on government policy measures does often not reach smaller firms. This is an area where general efforts to rectify the situation should prove helpful. On the other hand, closing information gaps on technological and economic/market development needs a more directed approach. Here the situation with respect to both available information and information needed differs considerably from industry to industry and even within an industry. Policy experiments should be undertaken on how best to assist different industries in cooperative efforts to gather, digest, and disseminate information.

Finance. In Germany, tax subsidies (special depreciation allowances or grants-in-aid on research-and-development investment) have proved to be of advantage to large corporations mainly. To support smaller firms 1) they need to be extended to all research-and-development expenditures including the manufacture of commercial prototypes, 2) the conditions for obtaining a subsidy have to be relaxed so that smaller firms find it easier to qualify, and 3) the money spent needs to be redirected toward smaller firms by providing funds only to firms up to a certain size or by introducing an absolute ceiling on the amount of research-and-development expenditure on which the subsidy is granted.

Supporting small firms through direct government grants out of programs to stimulate development in selected technology areas has proved very difficult. Major barriers are the nonavailability of information on these programs to smaller firms, the bureaucratic structure of ministries, the difficult and lengthy application process, and so on. Several efforts are underway to overcome these barriers (dissemination of a comprehensive information booklet, seminars on application procedures, special services to assist smaller firms in seletced technology fields), but quite probably the reluctance of smaller firms to apply for grants and the inability of the existing bureaucratic structure to cope with a considerable increase in the number of applications (and the necessity to decide fast) means that such efforts will show only limited success. To support and stimulate innovation in a larger number of small and medium-sized firms requires a policy instrument that is flexible enough to reach many firms briefly unbureaucratically. It seems that a measure of indirect support such as a tax subsidy geared specifically toward smaller firms will be most efficient. At least a policy experiment in this direction should be undertaken and its impact monitored.

Research and Development. Many small firms do not have the capability to engage in research and development and to innovate.

On the other hand, the scientific–technological infrastructure cannot substitute this missing capability. To support research and development in smaller companies, therefore, requires developing contract and industrial research institutes, technical information services, and so on, as well as assisting smaller firms to gain a minimal research-and-development capacity to make use of such services. Otherwise, mainly larger firms will benefit from them.

Cooperative research-and-development efforts among several small firms should be particularly initiated and supported. Where this is not feasible for competitive reasons, special efforts should be undertaken to stimulate cooperation between the supplier and the customer.

10 The Small High-Technology Firm

Peter R. Payne

I have no idea how many new, technology-based firms (NTBF) there are in the United States or what their survival rate is. Dollond et al.[1] say "several thousand" with sales running into "billions of dollars." This may be an understatement. The American Association of Small Research Companies alone has a mailing list of over 10,000, and its president, Dr. S. Cardon, feels this may be only a third of the total. But no one knows for sure.

The survival rate is also unknown. Cooper and Bruno suggest that it is much better than the average for all small companies based on a study of 250 NTBF's in the San Francisco peninsula.[2] But it's a truism in venture capital circles that one can easily find money for electronic-related ventures in northern California, much of it from the people who made so much money with computer-related companies in the 1950s and 1960s. It's like no other region in the United States and should not be extrapolated. In addition to that regional consideration, however, the field of electronics is a very special society, quite unlike the rest—an almost free, laissez faire economy, with sophisticated, performance-oriented customers. One cannot extrapolate its experience to other forms of NTBF's.

In the United States, NTBF's can be started with venture capital or by winning a government research-and-development contract, although the latter is becoming ever more difficult because contracting officers are demanding more stringent criteria for a company's financial resources. Like many other NTBF's, Payne, Inc. started in a basement with less than $500 capital and has never sold any equity.

I believe this ability to get started on a shoe string is unique to the

United States. One could certainly do it elsewhere in principle, but cultural and attitudinal factors within the society, coupled with very adverse tax structures, often prevent its happening in practice. In the United States, taxes are not that burdensome. A salary of $60,000 a year can mean $14,000 in federal and state income taxes, less if one works hard at tax sheltering. An entrepreneur can also have less visible, but important benefits such as a company car, combining vacations and business trips, etc. There's at least a chance his company will grow and eventually be worth a lot of money. On balance, it can be very rewarding, provided one has the intestinal fortitude to meet the payroll.

For the past 10 years, there has been a steady trend in the Department of Defense (DOD) toward inhibiting new start-ups and generally making life more difficult for those of us who are already in business. (As a result, Payne, Inc. no longer seeks defense business but is concentrating on the energy field, particularly nongovernment work.) This trend has already caused some problems for DOD itself, as noted by Deputy Assistant Secretary Gansler[3] in a perceptive review. I hope that Secretary Gansler and others will be able to reverse this trend, insofar as it stems from internal policies.

Congress passes laws that also make the smaller high technology companies' lives more difficult. It typically takes a year to get a contract instead of the 1 to 2 months that was the norm 10 years ago. The magnitude of the paperwork is almost unbelievable; yet in research-and-development work, at least, it's almost all meaningless. A 1-page work statement from the government engineer would be just as good. And in the Alice in Wonderland of competitive bidding, we sometimes charge the taxpayers $250 for every dollar let on contract in the process of throwing all this paper around.[4] Congress does not intend these results, but the actual results of a bill are often totally different from those intended.

For example, equal rights legislation is a major cause of minority group unemployment. Small businesses employ most of the people in the country. Since firing or laying off a minority group employee can cause a great deal of trouble because of provisions in the law, small businesses do not hire them in the first place.

Moving to the other method of starting a NTBF, venture capital, I find a picture that seems simple to me, complex to others. Lederman's perceptive paper suggests that "inadequate venture capital does not appear to be a basic problem in the U.S."[5] I would say that is corect in the sense that there is surplus money stored away. But, except for very special cases, it is almost impossible to convince the owners to invest it. These special cases are computer and communication-related ventures in northern California, typically requiring $500,000 to $1,000,000 for start-up, with more later, and nonnuclear energy and garbage processing, which are fashionable throughout the country.

The secret to venture capital investment is this. The investor would like to see a fledging company go public within 3 to 5 years so that he can recover both his money and a profit. With the present market psychology, it is almost impossible to be public.[6] Ergo, he will not invest.

The stock market psychology is not a reasonable thing, not susceptible to government control, except on the down side, and perhaps most closely related to America's self-image. We have recently been through a remarkably, I would say, absurd period of national self-flagellation and abasement, in which, to top everything off, we found that other countries could influence our lives and economy As a direct result, the small investors are out of the market, almost entirely, and the professional money managers motivated by "fear and greed" as someone has observed are not at all interested in new and unproved companies.

Where do we place the blame for the virtual nonavailability of venture capital? The answer is essentially with the whole of society, but especially perhaps the activist elements that have been trying to force social or political change of one kind or another. They have, in fact, caused some changes to take place, but not, perhaps, the kind of changes they looked for. As a result of these disturbances, there has been a drawing back by the majority, a trend toward conservatism. And, unfortunately for small companies, this conservatism has included withdrawing from the stock market.

In many ways, the vanguard of these revolutionary forces has been the media, which has also developed into an incredibly efficient feedback mechanism. An erroneous wire story about a sugar shortage on Monday can result in a real shortage by Friday. That is feedback! It seems clear that the gasoline crisis of 1974 was a similar phenomenon. Everyone suddenly wanted an average of 20 gallons instead of perhaps 5 gallons. That is, the supply system had to instantly produce over a billion gallons more than normal demand.

Notice that media feedback is invariably biased, and heavily so, in favor of bad news, gloom, or despondency. And, in the last two decades, we have seen the phenomenon of journalists with strong political or societal opinions practicing advocacy journalism, two words I never expected to see juxtaposed! So, in a sense, the media has become a cancer, destroying the society on which it feeds, by destroying its self-confidence.[7] This leads to a reduction in risk taking generally, and high-technology start-ups are among the first to suffer. I can see no way in which government action can ameliorate this significantly, except perhaps to encourage responsible journalism, as is done in the United Kingdom by the BBC, and leave the people to choose.

There are, however, some peripheral steps that might modestly improve the environment for NTBF's in the United States, a few of which follow.

As a practical matter, venture capital is not available. It may be a long time before our society changes its self-image and starts to increase its level of financial risk to the point where it does become available again.[8] But it would be helpful if we stopped the corrosive verbage that makes every successful entrepreneur a fat cat and equates profits with sin. Such demogogery can destroy us as surely as it destroyed Senatorial Rome. And populist political actions, such as the recently increased tax on capital gains, are not helpful either. I suspect they also reduce the revenues they were designed to increase, as Parkinson has suggested.[9]

Here are some specific suggestions for improvement that lean heavily on Morse's[10] outstanding analysis.

1. **Change capital gains tax.** Eliminate or greatly reduce capital gains tax rate for direct investment in NTBF's. Such an incentive should be available to both corporate and individual investors.

2. **Acquire founders' stock.** A new mechanism is needed to facilitate the acquisition of founders' stock by officers, directors, and key employees during the formative years of a company. Care should be taken to prevent adverse tax consequences that negate the value of the stock in attracting key talent to the enterprise team.

3. **Recognize the role of corporate investors.** The institutionalization of the venture capital community and the increasing use of the industrial corporate venture mechanism suggest that it would be desirable to allow corporate participation under both Subchapter S and Section 1244 of the Internal Revenue Code.

4. **Provide tax incentive for direct investment in small technical enterprises.** An immediate deduction against income for individual, institutional, and corporate investors for their direct investment in small technical enterprises would be an effective incentive for start-up financing. The investors would assume a

zero base, and capital gains tax liability would be incurred only upon sale of the investment.

5. Review SEC rules. SEC rules, notwithstanding Rule 144, continue to restrict the small company investor's liquidity. New combinations of holding periods and rates of distribution for both private and public companies should be considered.

6. Review tax regulations. General cost increases and inflation have made dollar limits in certain rules too small. The small business 22% tax rate should be applied to the first $100,000 of income rather than $25,000. The tax-loss carry-forward period should be extended from 5 to 10 years.

7. Review incentives for management. For the new enterprise, the value of stock options as a management incentive can be restored by reducing the holding period for shares issued under a qualified plan and by arranging to defer tax liability for shares issued under a nonqualified plan. Other forms of financial and tax incentives should be developed for the management and key employees of the higher-risk new technical enterprise.

Government Contracting with NTBF's

In the absence of venture capital, it seems we must maintain and improve government support of new technology. Historically, well over 90% of all new technology has originated in small companies, or has come from private individuals,[11] and the best way to get government support has always been the unsolicited proposal. Today, that is almost a dirty word in many branches of government. We need to change this thinking and to restore to the government engineers the power and prestige they had two decades ago. Government program managers should be encouraged to:

- Fund unsolicited proposals
- Expect a program failure ratio of 50% (no progress without risk)
- Get rid of all the planning nonsense that has grown up
- Use the Contracts Officers as servants rather than masters
- Be technically humble and remember that even Lord Rayleigh and Von Karman made mistakes in assessing new technology
- Go out on limbs and expect to be applauded, not reviled

- Speak out frankly about inefficiencies in the system and try to improve it

It would also be helpful if Congress got off its present management kick, where innovation has to be planned 10 to 15 years into the future, with the next 5 years pretty well defined. Planning innovative research is obviously as impossible as predicting new inventions. So nothing new gets funded, because it will not fit into the pre-existing plan.

Another problem with government funding is the business of actually getting it, after submission of a voucher. This can take months, years, even, in the case of excess overhead cost corrections. NTBF's typically do not have the financial clout to finance such delays. For small contracts, under $250,000, say, why not pay on a predetermined schedule, without the need to submit vouchers? This would solve the problem and allow hundreds of bureaucrats to move to more productive jobs than sorting, signing, transmitting (often losing) vouchers that are going to be paid automatically anyway.

Bank Loans

The Small Business Administration has established an excellent precedent for government-guaranteed loans, although they still have to be "good loans" in the bank's rather cautious judgment. We need to expand this sort of instrument so that it becomes a partial substitute for the equity capital that is no longer available.

Government Laboratories

With the space program winding down, there's a surplus of NASA engineers. With research-and-development budgets at their lowest point in two decades, almost all Department of Defense (DOD) laboratories are hungry for money as well. The result is intense competition for something that is funded, so that they can avoid lay-offs. And, of course, when selling to fellow bureaucrats, DOD labs have two outstanding advantages: They are all in the same family, and no contract is needed.

For example, the National Aeronautics and Space Administration (NASA) is heavily involved in solar energy. Of the Energy Research and Development Administration (ERDA) $17 × 10^6 windmill budget, $14 × 10^6 goes to NASA. ERDA's Sandia Labs has a stranglehold on parachute research in this country and is also spending millions on solar energy. The sole function of NASA's Energy and Enviromental Technology Office, as far as I can see, is to deflect money from various agencies into NASA. A sort of Civil Service WPA! I could go on and on with such examples. We have the danger that the labs will eventually soak up all the money appropriated by Congress for energy research, producing nothing with it.

In my view, none of the non-ERDA labs should be allowed to go outside their charters in this way. On top of that, we should carefully examine the function and past performance of all government labs and close down those that do not measure up. I can not believe we need the 834 separate labs identified in 1975[12] with 58 additional ones under construction and a total 1972 funding of $6.4 billion. And I know of some labs whose elimination would greatly improve the efficiency and productivity of the agency they work for, without any reference to the money saved. Perhaps the money saved could go to NTBF-related endeavors.

11 The Small High-Technology Professional Service Firm
Earl H. Hess

Lest we find ourselves in a dilemma similar to the group of blind men describing the elephant, this section should appropriately begin with a definition of terms, for a wide variety of small businesses have been described generally as "high-technology firms." The study by Dollond et al.[1] defined "new, technology-based firms" (NTBF) as those begun since 1950 and those set up by an individual or group of individuals for the purpose of exploiting an invention or technological innovation.

I limit my attention to an increasingly important sector of the United States economy—the professional services, small, high-technology firm. I describe what it is and what contribution it makes toward technological innovation and also discuss the climate in which it now functions within the United States. This type of firm consists of a group of scientists, engineers, and technicians who offer, for a fee, scientific and technological expertise in testing, analyzing, contracting research and development, and consulting in their fields of competence. Clients who purchase these services are individuals, large and small industrial organizations, government, and academic institutions.

These companies usually originate because of the desire of one or more technically trained persons to engage in private practice of their professions, analogous to the practice of medicine, law, and accounting. The start-up of such a private practice can take various forms and proceed in various directions. The business may begin and remain essentially a one-person operation, providing consultation, analysis, or other technical services in the field of the scientist's speciality. It is more typical for businesses of this type to begin as one- or two-person operations, but then to expand in response to their success in the development of markets for their services. Although there are numerous examples of such firms having grown within the span of one generation to multimillion dollar enterprises, the more typical high-technology professional services firm finds it advantageous to remain relatively small in order to retain its efficiency and flexibility, as well as the responsiveness and creativity encouraged by a close personal relationship with its clients. To my knowledge, there exists no accurate tally of the number of firms existing within the United States that would be correctly classified as small, high-technology professional services businesses. Directories of the associations representing such firms[2] are of some value, but one can never be certain as to what multiplier to apply to these membership lists in order to estimate total firms within the United States. The American Council of Independent Laboratories, for example, has a membership of about 200, and the American Association of Small Research Companies a membership of 250 to 300. Suffice it to say that there exists within the United States a viable industry of this type, that the role it plays in promoting technological innovation is a highly significant one (far out of proportion to its size as measured by sales or number of workers), and that the existence of such an industry is relatively unique to the United States economy.

The Industrial Climate in the United States

There are those who would suggest that the major reason for the existence of such an industry in the United States and not elsewhere (or to a very limited extent) is the "get rich quick" mentality of Americans that pervades the scientist's mind as well. Obviously a degree of financial success is essential for the continuity, growth, and vitality of any business, but there is a sharp distinction between sound financial planning and the greed often associated with independent businessman. In fact, from the very limited data available, it is apparent that many independently functioning scientists do not assess realistic fees for their services and therefore are considerably less financially successful than are their colleagues employed elsewhere.[3]

Thus it is obvious that reasons other than financial success are the prime motivators in establishing small, high-technology professional services enterprises. A need for greater independence and all that it entails, in my opinion, is the prime factor. Large institutions, whether in government, academia, or industry, tend to be highly structured to restrict of free movement and expression, and to foster a "cog-in-the-wheel" feeling in individuals (i.e., a feeling of frustration caused by the inability to change significantly the nature of their jobs). On the other hand, there are those of us who feel strongly enough about the matter of independence in our profession to risk everything to attain it. Other reasons include the desire to apply scientific talents to a broader field, the challenge of acquiring business skills, and the pursuit of a career that combines both technical and business activities.

I have dealt thus far only with the factors that would seem to motivate scientists to establish independent professional services firms, and I doubt that these characteristics are unique to American scientists. Motivation to begin, however, is only the first step. A need for the services offered and a business environment conducive to survival and growth are equally essential elements. It is in these latter respects that the United States would seem to be unique.

The success of the American economic system in the past has been generally attributed to its free enterprise system, wherein private business has been allowed to function with minimal government involvement, in contrast to highly socialized systems where all kinds of services flow from the government, made possible by high taxation. Looking at American industry more closely, we can see an intricate inner network of individual companies acting as basic commodity suppliers, fabricators, and marketers, all of which are interdependent and collectively and effectively provide the nation's goods and services. Small business is a vital part of this whole network. Specifically, scientists who have established small professional services firms have found their place within this system working in cooperation with, in support of, and sometimes in place of scientists and engineers from within American industry. In private industry the "make or buy" decision is usually made based on one simple factor, namely cost-effectiveness.

The innovative productivity of this segment of the scientific community has consistently exceeded that of larger, more structured organizations using the "research by committee" approach.[4] Penicillin, the jet engine, DDT, Dacron polyester fiber, and the Polaroid camera represent five examples from a long list of highly significant technological innovations accomplished by small companies or individual inventors. In making a subtle but significant distinction between managers and leaders, A. Zaleznik provides indirectly one of the best explanations for this superior performance.[5] Leaders and managers, he contends, "are very different kinds of people. They differ in motivation, personal history, and in how they think and act." His description of the conditions within our present bureaucratic society that "breed managers" and "stifle young leaders" is sobering indeed.

Everyone with experience in the development of new products or processes is aware of the infinite number of small problems that must be solved after the major technical breakthrough and before commercialization. Small, independent, high-technology firms play a very useful support role to American industry in this area. The development of analytical methodologies, the design of quality control systems, the performance of environmental impact studies, developing means of satisfying government regulations, finding creative uses for by-products, and performing quality control analyses are but a few examples of this significant function.

Although the independent laboratory plays a valued role in support of the innovation process for large American industries, its services are probably even more vital to small manufacturing businesses. The basic changes undergone by our society in recent years—advancing technology, the consumer movement, sensitivity to environmental impacts of technology, and the consequent avalanche of government regulations—require a degree of technical competence in almost all busineses. It is usually more cost-effective for small businesses to acquire that competence through purchasing professional services than by generating an in-house capability.

It is more cost-effective for both large and small business to use our industry than to provide an in-house capability, although the argument could be made that government could provide the same support more effectively. This is, of course, the classic socialist response to the free market philosophy and suffers from the basic defect of socialism, that is, that in the real world things just do not work out that simply. One specific example of such an attempt by the United States government illustrates my point. The cooperative extension service of the United States Department of Agriculture is a huge organization, costing the taxpayer heavily, that in effect provides a socialized technical service to agriculture. In practically every heavily argicultural area, there exists a viable independent laboratory community providing, for a fee, many of the same kinds of services provided at little or no charge by the extension service.

When an industry can survive under such circumstances with its own taxes subsidizing its competition, then indeed its efficiency and quality of work must exceed its competition by at least an order of magnitude.

In summary, government and academia serve very useful purposes in providing a broad base of basic scientific information. However, technological innovation must come from private industry, because only there is the nature and magnitude of the technological problems and economic constraints on solutions of these problems fully appreciated. Small, high-technology professional services firms find a significant place in the technological innovation process because of their unique qualifications for the job. Their services *must* be of value to their clients, or they will not survive.

The Future Outlook

What then, one might ask, does the future hold for small, high-technology professional services firms in the United States? For indeed, the reality of a growing socialism is painfully obvious to all in the private sector. Unfortunately, the impact of socialism is felt first and most dramatically by small and young businesses. The burden of an unemployment compensation program that has been converted to a welfare handout, of the inequitable burden of social security premiums on the principals in a small closed corporation, the complexities and inequities of estate taxes that often force families to relinquish control of a business upon the death of the founder, the tremendous burden of the governmental red tape[6] serve to illustrate the realities faced by the small businessman of the 1970s.

Then too there is a series of government activities that impact very specifically in a negative sense on small, high-technology service firms. Various government agencies have developed research-and-development capabilities far in excess of in-house needs. Such excess capabilities are seldom dismantled or converted when their original mission is accomplished, but emerge as direct competitors to small private enterprise firms for government contracts. The National Aeronautics and Space Administration (NASA) is a prime example of such a case, having turned from space research to the performance of contract research for agencies such as the Energy Research and Development Administration (ERDA). A comprehensive study of federal laboratories, their management, and utilization reached conclusions most uncomplimentary to them.[7]

Direct and unfair competition by tax-favored (so-called "not for profit") research institutions and universities in the performance of strictly commercial work is a growing problem for the private sector. The nature and extent of this problem is addressed by a recent memorandum produced for ACIL.[8] Independent laboratories are now seeking assistance from the office of advocacy of the Small Business Administration (SBA) so that agencies of the federal government might be aware that private, tax-paying service firms should not be threatened by unfair competition from tax-supported institutions.

Procurement policies and protocols for federal government contract work have become so complex that many small businesses have become so complex that many small businesses have simply gotten out of the business. The red tape of Equal Opportunity Employment Certifications, the rigidity of allowable cost accounting procedures, the weight assigned to previous government contracting experience, all act against the young or small research-and-development company.

Then too there are areas in which small-high-technology service firms have sorely needed government help to preserve a climate conducive to their survival, yet have not found it. For example, the consumer movement, certainly supported in large part by the government through vote-conscious politicians has had some far-reaching side effects, one being the matter of product and professional liability. Costs for such insurance protection have either skyrocketed or the coverage has been withdrawn. Most small professional service firms lack any kind of protection, so that one's life savings are actually on the line with each $15 analysis reported.

The independent laboratory community has emphasized the need for a commonly accepted procedure for recognizing professional competency. It has developed its own set of standards for assessing the qualifications of laboratories[9] and has pressed government to assist in establishing a competent, disinterested third party agency to acredit laboratories. Unfortunately any positive response from the government has been slow in coming, unnecessarily complicated, bureaucratic, and not at all unified among agencies within the government. Therefore, even though some laboratory work proceeds unregulated as to quality, other laboratory managers may find themselves encumbered by a web of complicated and overlapping accreditations that strangles their productivity. It can be hoped that efforts presently being made by the United States Department of Commerce, National Bureau of Standards will bring some order to this chaotic situation.

The Small High-Technology Professional Service Firm 117

Of course, there is a brighter side than that described in the last few paragraphs. Although a definite trend toward greater government regulation exists, the private enterprise system is not in real jeopardy. Small, high-technology professional service firms will continue to find their place in that system. Additionally, government has been taking a more serious look at the role of small business in general and small, high-technology firms in particular. The prospects of the elevation of the SBA to departmental status is encouraging. The existence of the National Science Foundation's Research Applied to National Needs (RANN) program, "small business set asides" in the government's procurement of professional services, and new tax legislation favorable to small business, are ample evidence of the government's concern for this vital sector of the economy. I am further encouraged by the special attention given to small business by ERDA, which is currently establishing dialogue with small business, and simplifying procurement processes. The inclusion of a representative of small, high-technology business on this program is another example of the growing recognition of this section of American industry.

Summary

A significant number of viable small, high-technology professional services firms presently exist in the United States because they play a significant role in the free enterprise system. The unstructured environment of the small firm seems more conducive to technological innovation than does the more highly organized climate of larger organizations. Further, the flexibility and adaptability of small high-technology professional services firms permit them to offer a variety of support services to the technical innovator on the most cost-effective basis. Trends toward socialism and excessive government involvement and regulation threaten the existence of this industry unless conscious effort is made to soften the negative impact of such actions.

PART THREE

GOVERNMENT: POLICIES AND PRACTICES FOR INDUSTRIAL INNOVATION

12 Government Policy and Innovation in England

Michael Gibbons
P. J. Gummett

One of the most perplexing aspects of current thinking about economic growth concerns the role played by technical progress in the process. The perplexity arises because, for a long time, technical change was regarded as an exogenous variable, its contribution to growth appearing under the notion of the residual in most growth accounting schemes. By contrast, now it is recognized widely that technical progress, or technical change, constitutes a deliberate choice, involving allocation of resources and, as such, should be included along with other investment choices as an integral part of the theory of the firm. The identification of technical change as a deliberate option in the dynamics of the firm is at least partly traceable to the recent emergence of research and development as an independent functional activity of firms. As Freeman has pointed out,[1] these programs play diverse roles in overall corporate strategy, depending upon whether that strategy is expansionist, contractionist, or oriented to maintenance of the status quo.

Quite apart from the de facto role played by research and development in the corporate life of individual firms, there persists a widespread belief that research and development is essential for the maintenance of a technically progressive industrial sector. One consequence of this has been the increasing government involvement, since World War II, in research-and-development activities, extending from the funding of basic research in universities to the development of new research-intensive industries such as aerospace and nuclear power. The idea of research and development as a national resource was accepted, perhaps somewhat uncritically, as a remedy for a wide range of industrial problems not the least in Britain, where the fusion of technology and socialism was expected to provide fuel for a "white hot technological revolution" that was to revitalize British industry. When research and development failed to deliver the goods—indeed, it could not have lived up to all that was expected of it—many national governments began to rethink their research policies. Essentially, each asked the same question, "What relationship is there, if any, to the existence of a research capability and the achievements of national goals." Not surprisingly, each country tackled the problem somewhat differently—some more imaginatively than others[2]—but the shape of the answer was that the relationship was more complex than is commonly supposed and that new administrative structures, design to keep a closer eye on public spending on research and development, were necessary. In brief, decision making with respect to research and development was to be opened to greater public scrutiny and made more accountable.

The first part of this paper describes one aspect of the attempt made in Britain between 1970 and 1977 to orient government-sponsored civilian research and development more in the direction of national needs through the mechanism of the customer–contractor principle prescribed by Lord Rothschild.[3] We shall be concerned mostly with the application of that principle to the Department of Industry (DOI) and its Industrial Research Establishments (IREs). It is important to realize, however, that although the Rothschild reforms refer mainly to administrative procedures in the DOI, the particular form that these procedures took has changed significantly both the locus of decision making and the capacity of the department to direct research to specific ends.

From the perspective of the economist interested in the relationship between technical change and growth, or the policy analyst interested in the process of technological innovation and its stimulation, it is important to ask what effect, if any, these administrative reforms are likely to have on innovative performance of specific firms and industries. It is possible, theoretically, for the machinery for implementing the customer–contractor principle to function smoothly and efficiently and still leave the behavior of individual firms unchanged. The second part of this paper examines some of the implications of the Rothschild reforms for the context of innovation in British industry.

The Department of Industry

The new arangements for administering research in the DOI, and its predecessor the Department of Trade and Industry (DTI), have been described elsewhere in more detail than is possible here.[4] The Department of Trade Industry was formed in 1970 from the Ministry of Technology and the Board of Trade, and assumed responsibility for several so-called industrial research establishments such as the National Physical Laboratory (NPL) and the National Engineering Laboratory (NEL) together with the United Kingdom Atomic Energy Authority (AEA) and several joint government–industry research associations. In January 1974, the Department of Energy was created to assume part of DTI's responsibilities, including the AEA, and in October two other departments were separated out, leaving a reduced Department of Industry. The DOI, however, continued to have responsibility for nonnuclear research conducted in AEA establishments, particularly Harwell.

As early as 1969, Mintech, and later DTI, began to review the work of the IREs by program area rather than by laboratory,[5] feeling that this would be more systematic in a climate more favorable to programmatic approaches to policy making. Moves were made to define programs in terms of their end users, and one of DTI's permanent secretaries later dated the end of 1970 as the moment when DTI felt able in formulating research programs "to rely less on general advice from the scientific community . . . and more on a structure aimed at identifying and involving all the end users."[6] That is, even before the Rothschild report, DTI had been moving toward some form of customer–contractor arrangement for formulating research policy, and by March 1972, only a few months after the publication of the report, the department presented an outline of its proposed new structures to the Select Committee on Science and Technology.[7]

The White Paper that followed the Rothschild report in July 1972, announced that DTI was establishing a set of Research Requirements Boards (to play the Rothschild customer role) supported by a Research Requirements Division and matched (for the purposes of Rothschild's Conrtoller of research-and development function) by a Research Contractors Division.[8] Boards were quickly established to oversee the following areas: standards and metrology; ship and marine technology; chemical and mineral processes and plant; computers, systems, and electronics; and engineering materials.[9] The boards comprise civil servants, industrialists, and academics and are not merely advisory but have the power to spend a defined budget in IREs, AEA establshments, research associations, universities, industry, and research councils. From the point of view of the IRE's, the chief potential effect of the new structure was described by Dr. Maddock (now Sir Ieuan), the the Department's Chief Scientist.[10]

"Rather than the present situation where the program is more or less invented with the IREs and then endorsed by a process of advisory committees and ultimately by myself, the programs will be arrived at by bringing together different requirements which have been examined by competent boards."

Each laboratory's program, in other words, should comprise a portfolio of requirements from several boards, the focus of interest shifting toward the overall cluster of requirements issued by each board and away from the overall programs of individual laboratories.

Not all research and development conducted by the IREs, or within DTI, was to be submitted to the boards for their approval. Work carried out by IREs for industry on full repayment of costs would be outside the remits of the boards, although the latter would be kept informed. Furthermore, when DTI supported extramural research and development with the primary objective of providing direct assistance to industry, and the research-and-development aspect was secondary, the approval of a Requirements Board would be unnecessary. In effect, therefore, all the high technology supported by DTI, such as aerospace and nuclear power development, was to lie outside the Requirements Board system, leaving the latter, in late 1972, responsible for some £12 million of research-and-development programs, or about 7% of DTI's total expenditure. (The proportion will be higher now that DOI is no longer responsible for the AEA.) Only a small proportion of DTI's total research-and-development spending, this sum was, nevertheless, highly significant for the IRE's and the nonnuclear work of the AEA laboratories.

The Requirements Boards, unlike the comparable bodies in the Department of the Environment (DOE) and Ministry of Agriculture, Fisheries, and Food (MAFF), regard themselves not as real customers for the work of the IREs but as proxy customers.[11] It is argued that, unlike DOE and MAFF, which can themselves use research results in formulating policies and framing regulations, DOI has very little use itself for research and development and is instead seeking to obtain research results that will be of value to British industry. Hence, unlike the DOE and MAFF cases, industrialists are appointed to the decision-making bodies. Not surprisingly, the role of proxy customer

Government Policy and Innovation In England   125

is hard to play, not least because it is not obvious what kinds of research programs will most benefit British industry.

This question could not be systematically addressed by the boards in the first 12 to 18 months of their existence because, from their inception, the IREs technically had no customers, and the boards were faced with the task of rapidly processing and reviewing applications for a set of programs that already were underway in full spate. This is not so say, however, that in their haste to implement the new arrangements, the boards were uncritical of existing programs. As one official put it, many submissions had a "very, very rough ride."[12] By the end of 1973, however, most boards were well advanced in their reviews of their areas and began to make serious efforts to clarify their criteria for choice of projects. A document discussing the resultant interim strategies of the boards was published in early 1975,[13] and showed that although the task of defining strategies was not easy, a good start had at least been made.[14]

The Composition and Operations of the Boards

A seventh board was created to be concerned with Fundamental Standards and an eighth was called the Chief Scientist's Requirements Board. The former proved to be a temporary measure and its work was transferred to the Metrology and Standards Board early in 1975. The Chief Scientist's Board draws ad hoc upon a pool of potential members that includes some chairmen and members of other boards and considers matters that do not fall readily to any other board. In late 1975, a new Garments and Allied Industries Requirements Board was set up to assume responsibility for an area that had been handled previously by the Chief Scientist's Board.

One final note on machinery is that general responsibility for the operation of the requirements board system rests with the chief scientist in the department. He oversees the boards and brings a wider viewpoint to bear on their work through meetings of a committee of chairmen of requirements boards. This committee is considered, within the department, to be a valuable guard against potential parochialism within the boards and also provides a forum for deliberations over matters of common policy.

In 1972–1973 and 1973–1974, four of the six boards were under the chairmanship of industrialists and about half the members of each board were industrialists (Table 12–1). This was in accord with

Table 12-1. Composition of Department of Industry Requirements Boards, 1972–1973/1973–1974

Board	Industrialists 1972–1973	Industrialists 1973–1974	Academics 1972–1973	Academics 1973–1974	Officials 1972–1973	Officials 1973–1974	Total 1972–1973	Total 1973–1974
Ship and Marine Technology	8[a]	8[a]	1	1	5	6	14	15
Chemical and Mineral Processes	8	10	2	2	4[a]	6[a]	14	18
Engineering Materials	6[a]	6[a]	1	2	4	4	11	12
Mechanical Engineering and Machine Tools	5[a]	6[a]	2	2	4	4	11	12
Computers, Systems, and Electronics	7	7	2[b]	2	5[a]	5[a]	14	14
Metrology and Standards	6[a]	7[a]	2	2	5	5	13	14

[a] The number includes the chairman.
[b] The Director of the Royal Greenwich Observatory is included.

Source: Department of Industry, *Report on Research and Development 1972–73* and *Report on Research and Development 1974*, (London: M. Stationery Office, 1973 and 1975), pp. 44–49, 108–114.

Government Policy and Innovation In England 127

the department's intention of giving "a strong voice to industry to ensure that the work of the Boards support will be relevant to user needs."[15]

Great care is taken in selecting the industrial members. As one official observed, the boards are "handpicked . . . the last thing we wanted was just a bunch of research enthusiasts." Another official gave as an important criterion in selecting industrial members that they be "at the point where industry makes its decisions," that is, at, or near, board level. This, presumably, partially accounts for the conspicuous absence of trade unionists from the boards. A further criterion was that the members should be able to see the relations of technology to industry. Therefore, they might be research managers but were likely, also, to be marketing managers; they were unlikely to be accountants. The calibre of the industrialists selected looks, certainty on paper, as high as reasonably might have been hoped. Of the 40 industrialists on the boards in 1972–1973, 12 were described as managing directors or assistant managing directors, seven as chairmen or vice-chairmen, and 12 as senior research or technical directors. The list of companies from which they came included BP, the British Steel Corporation, Ferranti, GKN, ICI, IMI, Joseph Lucas, Laporte, Marks and Spencer, Rank Research, RTZ, Rolls-Royce, and The Weir Group.

Of the officials on the boards, at least one, and usually two, come, not from the headquarters office of the Department of Industry, but from the industrial research establishments or from Harwell. This, together with frequent appearances before the boards of staff from the establishments, who come to support their submissions to the boards, has led to a high level of interchange between the research establishments, headquarters staff, and the industrialists on the boards, leading the department to claim this interaction "between practitioners and administrators had brought good dividends."[16] One of these dividends appears to have been a real, and not merely cosmetic, improvement in the quality of submissions from establishments to boards.[17]

Some liaison between the boards and the industry divisions of the department is maintained by those officials who are responsible for serving the boards and the research establishments. Organizationally, the new procedures mean that the Research Division of the department was divided into a Research Requirements Division, serving the establishments. It is worth noting here that, even though relations between the establishments and the Research Contractors Division are said to be excellent, there exists some feeling in the establishments that Research Requirements staff are "them," rather than "us." This is not necessarily undesirable but is a sign that the boards and their staff are regarded with caution by staff in the research establishments.

The Strategies of the Boards

By the end of 1973, most of the boards were well advanved in their reviews of their areas, and all were beginning to pay serious attention to their future strategies. Thus the Chief Scientist's Board observed that it had tried consistently to establish a "need-oriented" framework within which to make judgments. This had involved such devices as setting targets for cost recovery from industry to provide evidence of demand, setting firm time limits on departmental support, and identifying immediate uses of the program and plans for disseminating its results. But the time had come to try to make explicit the criteria against which funds should be allocated.[18] Another board, Engineering Materials, observed that almost all the programs so far considered were initiated by contractor organizations, that is, chiefly, the department's research establishments and the research associations. Although the board expected this in the early stages of such a new venture as the requirements boards, the board believed, nevertheless, that it was "beginning to acquire the necessary experience to take more initiative in influencing the balance of effort in its field."[19]

In 1974, therefore, about 18 months after their inception, the boards began to make serious efforts to take time out in special meetings, and even at dinner parties, to try to clarify their criteria for choice. The Minister had held meetings of chairmen for this purpose since the inception of the boards, but now the pace was increased. This proved to be a more difficult task than had been expected, but in early 1975 the boards were able to publish a document outlining their interim strategies. In his foreword to this document, Sir Ieuan Maddock observed that the boards were beginning to develop criteria for the deployment of funds "which need not necessarily imply an extrapolation into the future [of] the needs of the past." These strategies, however, should be regarded as tentative, and the purpose of publishing them was to "initiate an expanding dialogue" between the boards and industry so that policy would be broadly based and well informed.[20]

Despite the differences dictated by their varying interests, certain objectives and criteria were held in common by all the boards. Thus all boards took it as axiomatic that the work they funded was not an end in itself and that it must bring ultimate benefits, whether quantifiable or not. Development of expertise in particular techniques was was, therefore, a justifiable objective for them; what mattered was the use to be made of the technique and the likely consequences of that use. Another general objective was to improve the efficiency and international competitiveness of British industry and, as an important aspect of this, to improve Britain's balance of payments, whether by increased exports, greater added value to manufactures, substitution, or whatever. Furthermore, all boards had to consider the relative priorities of work aimed at commercial benefits to the industry concerned and work intended "in the broadest sense to improve the quality of life." A further problem was to decide whether a project, however desirable intrinsically, was an appropriate subject for government support, or whether it should be conducted by the industry itself. All boards also tried to ensure that a mechanism for technology transfer was incorporated into research proposals.[21]

We examine chiefly the work of two boards: the Chief Scientist's and the Computers, Systems, and Electronics Requirements Boards. Their rather different approaches give some idea of the range and depth of thought exhibited by the boards. The Chief Scientist's Requirements Board oversees a miscellany of research areas that have in common only the fact that no other board can appropriately consider them. These have included, in particular, projects arising from research associations in traditional sectors of industry such as textiles, shoes, and pottery. Even though it had not been possible to develop a unified approach toward all these projects, some common elements had emerged from the board's deliberations. These elements included the attempt to identify the potential benefits and beneficiaries of each project, assessments of the likelihood that these benefits would, in fact, be obtained using revenue from repayment as a measure of the value of the work, and the consideration of why government and, in particular, the Department of Industry, should fund the research and development.[22] No details were given, however, of precisely how these tasks were performd.

The board found that many of the proposals from research associations were of a type that lent them to a strategic mode of assessment. The meaning attached to this term is grasped best by illustration. A number of proposals had been received from the British Launderers', Rubber and Plastics, and other research associations.

The proposals were considered in four stages:

1. Examination of the future prospects for the industrial sector(s) served by the [research association].
2. Assessment of the benefits to the sector of additional R and D being undertaken, over and above that done by the industry itself.
3. Judgment of what constitutes an important industrial sector and, therefore, reasons for government support for the sector through funding R and D.... The primary criterion for support should be the impact on the balance of trade.... Import substitution, either directly or by recycling, more efficient uses of imported raw materials, and increasing value added to raw materials, were considered very important. Also important, though of lower priority, was improvement of the noneconomic quality of life through research aimed at the protection of health, consumer protection, and pollution control.... It was agreed that the Department of Industry could provide assistance to industry through R and D to meet regulations in these areas—if the long-term prospects of the industry warranted it.... Industries which were fragmented, making low profits, or too conservative in assessing the benefits available from R and D, should only be supported if their long-term prospects were good and if the support did not hinder rationalization ... into more economically viable units. Finally, the relationship between the various sectors of industry was noted since the well-being of one sector could be ... of vital importance to another.
4. Suitability of the contract.

Whereas stages 1 and 2 involve some difficult assessments, and stage 4 involves a judgment about the appropriateness of the contractor, it is with stage 3 that we approach the heart of the policy problems that arise in this field. It is worth noting here the prime significance, in deciding whether the department should support a project, of the balance of trade, and the secondary importance of improving the industrial provision for the protection of health and the consumer, and for controling pollution. The extent to which this ranking reflects the opinion of departmental officials as opposed to industrial members cannot be gauged. Probably, it is a mixture of both and, although this is an unsurprising ranking to emerge from a body such as the Chief Scientist's Requirements Board, stronger signs that the board is prepared to question the less desirable effects

Government Policy and Innovation In England   131

of industrial practice, and that it is not prepared to lave the initiative in such matters primarily with other departments of government, would be welcome. Also noteworthy under stage 3 is the requirement to examine the long-term prospects of fragmented or noninnovative industries prior to offering them support. Exactly how this is to be done, however, is not made clear.

If the Chief Scientist's Board were concerned primarily with considering why it should support any particular industry, the Computers, Systems, and Electronic Board's central question was why it should support any particular proposal within its more clearly defined area of interest. That is, the Computers, Systems, and Electronic Board has less need than the Chief Scientist's Board to consider the long-term prospects of different industries. On the other hand, because its task is to allocate funds within a defined technological field, it may have more need of a finely detailed set of criteria for assessing the relative merits of proposals.

Consequently, while the strategic statement of the Computers, Systems, and Electronic Board considered at some length the general reasons for government support of research, the central part of its statement was a list of criteria for support of research projects. These were:

Relevance. To what extent does the project serve either: Departmental responsibility in the Board's field, or a widespread need for which the Board is a proxy customer or some other purpose?

Necessity. Could the same end be achieved by other means without the proposed R and D?

Novelty. Are the results not already avalable or likely to become available in the United Kingdom or elsewhere? Has the Department political, economic or commercial reasons for not buying in the information or technology?

Credibility. Is the R and D clearly specified and likely to achieve its object in the appropriate time-scale?

Exploitability. Will the results of the work be used for the purpose planned? Do the necessary conditions exist for this? Can they be established? Do competent firms, capital, manpower facilities or legislation exist?

Cost effectiveness. Is the return to the economy sufficient to justify the cost of the project to the Department?

Dependence on support. Without government support would the work not be done by private initiative?

Competitiveness. Taking into account the effect on overseas competition, both negative and positive, does the outcome provide a nett [sic] benefit to the United Kingdom?

Type of research. Is the work curiosity— or truly application—oriented? Where noncommercial considerations apply, other criteria may be involved.[23]

In addition, it was necessary to consider the appropriateness of particular contractors for each type of project and not to concentrate solely on those establishments for which the Department of Industry was responsible.

Not all the boards felt able to say a great deal about the evolution of their strategies, and some had developed their thinking further than others. If, in the words of one board chairman, the interim strategies were still "rather naive and a little platitudinous," they did nevertheless, set in train a style of thought that has been pursued.

One example of this relates to the emphasis placed by the boards on technology transfer. Various devices have been tried to ensure that new knowledge is appreciated by potential users and that existing knowledge is disseminated adequately. One route to this end has been to publicize the existence of the boards and the programs of the boards themselves as important links in the national flow of technological information. Thus since 1973, the chairman and members of the Ship and Marine Technology Requirements Board have been active in visiting industrial companies to talk over matters of common interests, and in July 1973, the same board wrote to 263 organizations and companies about the content of its program. Replies were received from 160 companies, and the major reactions, the board reported,

> were to endorse taking a central view across the field of ship and marine technology and to welcome with enthusiasm the invitation to comment; while there was a widely held opinion that resources might be too thinly spread no one suggested the omission of any items; on the contrary some additions were suggested.[24]

The secretariat of the board followed up some of these suggestions with individual companies. A similar exercise was conducted a year later, and both in that year and in 1975–1976, seminars were held on aspects of the board's program. In addition, in 1976, the chairman issued his own report on the board's work to a large number of organizations and firms.[25]

Government Policy and Innovation In England   133

The boards are involved closely in the support of the industrial research establishments, their principal source of funds, and it is to the effects of the boards on those establishments that we turn now.

The Industrial Research Establishments

The requirements boards were not required to safeguard the laboratories for which DTI was responsible. Their objective, primarily, was to identify what research and development the department should support and where that work could be done best. Maintaining appropriate balances between and within laboratories was to be the concern of the Research Contractors Division within the department. Nevertheless, as one board has noted, government research establishments do constitute "an important national resource and a massive going concern. It is, therefore, difficult to impose an entirely new strategy on them, it has to be moulded from their ongoing interests and programs."[26] The same board added, however, that changes in laboratory programs could be expected, with likely effects on recruitment policies and future thinking within each laboratory.

Dr. Maddock had used blunter language in 1972. He described the situation at that time as one in which each laboratory director was "very much captain of his own ship" and had "a great deal of independence."[27] In the future, each establishment would end up with a "portfolio" of requirements that would have arisen through different boards.[28]

In 1971–1972, DTI spent £13.1 million in six industrial research establishments (Table 12-2). Torry Research Station was transferred to MAFF in April 1972, and the Safety in Mines Research Establishment to the Department of Energy in 1974 and, later, to the Health and Safety Executive of the Department of Employment. The principal establishments for our purposes, therefore, are the NPL, the NEL, and Warren Spring Laboratory. It is worth noting at this stage that the last-named differs from the other two in that a large proportion of its funds, about 50%, come from the Department of the Environment and are, therefore, outside the control of the boards.

The requirements board system has represented a change toward stronger control of the research establishments, a change to which the establishments seem, by and large, to have accommodated themselves. There is more appreciation now of what is required of them by the boards, although some feeling exists still that they would like clearer guidelines from the boards. Whereas project leaders formerly

Table 12-2. Expenditure by the Department of Trade and Industry on the Industrial Research Establishments, 1971–1972

	£ million
National Physical Laboratory	5.6
National Engineering Laboratory	3.3
Warren Spring Laboratory	1.5
Laboratory of the Government Chemist	1.0
Safeyt in Mines Research Establishment	1.1
Torry Research Station	0.6
Total	13.1

Source: Select Committee on Science and Technology, Session 1971–72, *Research and Development Minutes of Evidence and Appendices*, HC 375 (London: H. M. Stationery Office, 1972), p. 330.

would have sought support from their seniors for a project that seemed simply a good idea, now they accompany their arguments with details of costs, benefits, likely markets, related work being done elsewhere, and so forth.

One disadvantage of the new system is that it is time consuming and hence financially costly. Certainly, a good deal more staff time seems now to be spent in committees and other work preparatory to obtaining support from a board. There has been also an increase in paperwork, although, as one official commented, this can be seen as a "useful, if sometimes irritating discipline."

There can be a degree of inflexibility in the speed with which submissions are processed, and this can lead to difficulties when an establishment is seeking to marry a potential industrial contact with board support. Also difficulties can arise when a board shows insufficient awareness, not only of the awarkness of negotiating contracts under these conditions but also of the time that an industrial customer may take to pay over his share of the costs of a project. Although there may be some truth in the latter claim, and no doubt instances of the former difficulty can be cited, in general the system for processing submissionsn seems now to proceed quite smoothly. The boards felt able to claim, in 1976, that they "are willing and able to take special steps to ensure speedy decisions are taken, as required, to get the first stages of a project started without delay."[29] Implicit in that claim is the fact that officials in the Research Requirements Division have been authorized by the boards to act on their behalf, within the limits of their delegated authority.

The Context of Innovation: Some Concluding Observations

The first part of this paper described the genesis and operation of the Research Requirements Boards and investigated, tentatively, their effects to date on DOI's research establishments. The discussion thus far has been about the administrative arrangements adopted by DOI in its attempt to implement the customer–contractor principle as a vehicle for ensuring that government-sponsored research programs are tied to departmental objectives. As indicated earlier, the organizational changes have been considerable, and, consequently, it is too early for a full evaluation of them. Nonetheless, it is possible to make some preliminary observations about the likely impact of these changes on what we have called the context of industrial innovation in Britain.

The first observation refers to the role of DOI as a proxy customer for industry in relation to research and development. Unlike DOE and MAFF, for example, DOI has no direct use for the output of research from its various establishments. What research is performed is intended to meet the needs of industry. But what are the needs of industry? Here, we approach the theoretical core of the customer–contractor principle: In general, the departments as customers are supposed to place contracts for research that they deem relevant to their objectives. It is presupposed that these departments know, can articulate, what they want and that the laboratories can respond by means of feasible research projects. This notion would seem to be based on a highly simplified view of the relationship that exists between a customer and a supplier in a traditional market-oriented business environment: The customer knows what he wants and is willing to pay a certain price for the commodity.

Even though one may have reservations about the appropriateness of regarding research as a commodity, the customer–contractor principle itself would seem to be strained further in the context of DOI. If the department itself has no need for research per se, how are the research programs to be developed? Strictly speaking, industry has no needs, though sectors and firms do, but it is difficult to know where the relevant knowledge lies. To economists, there is nothing new about this problem. Briefly, it is but one aspect of the familiar dilemma concerning the relationship between microeconomic phenomena of growth.[30] This problem is posed nicely in a recent article by Nordhaus and Tobin:

> The [neoclassical] theory conceals, either in aggregation or in the abstract generality of multisector models, all the drama of the events—the rise and fall of products, technologies and industries, and the accompanying transformation of the spatial and occupational distribution of the population. Many economists agree with the broad outlines of Schumpeter's vision of capitalist development, which is a far cry from the growth models made nowadays in either Cambridge, Mass., or Cambridge, England. But visions of that kind have yet to be transformed into theory that can be applied to every day analytical empirical work.[31]

As Nordhaus and Tobin suggests, "the drama of events" takes place at the level of the firms. How the behavior of individual firms in pursuit of innovation can be related to corporate research and development strategies is a problem that is exercising economists a great deal at the moment.[32] The further problem of trying to aggregate individual strategies into sector strategies and, hence, sector needs has, to our knowledge, received very little attention so far. Indeed, it may not be necessary even to attempt it. For, if Schumpeter is right, it is precisely the nexus of need and technological opportunity that provides the basis for "entrepreneurial profit" that, in turn, carries the economy from one technological regime to another.

Certainly, DOI was aware of its role as a proxy-customer for industry when it established the research requirements system. But does an industrial presence in the boards meet the problem effectively? What indications are there that these representatives, with all the good will in the world, will be able to take the broader view? At present, it is too early to say whether the boards will overcome the problem of aggregation and actually be able to prescribe research of a sufficiently general nature to meet needs and problems at the sector level. The only evidence we have at the moment is that some within the research establishment are being given a very rough ride indeed and that the boards are beginning to play a more positive role in structuring laboratory programs. With regard to the latter, many of the boards have produced a list of the criteria against which they propose to select projects for support. In general, the criteria aim to meet sector needs, but still one is left in the dark about how these needs are to be identified. To the degree that the links between needs and research and development remain unclassified, one must

Government Policy and Innovation In England 137

remain skeptical as to the impact of the Rothschild reforms on the innovative performance of British industry.

The second observation begins from a very different viewpoint. It is precisely because technological innovation is known to be a complex and uncertain process that governments have sought to build a sound infrastructure of scientific and technological services. Indeed, most of the current literature on technological innovation indicates the wide variety of knowledge links that individual firms may have to the matrix of scientific and technical knowledge, both national and international.[33] Of course, the argument is strongest in the case of basic research, but there is plenty of evidence to show the important role played by government research establishments and research associations in the genesis of technological innovation.[34] From this perspective it can be asked if it is possible that the research requirement boards will focus too narrowly on short-term applied research thus decreasing the real range of options in the longer term? Centralizing decision making in relation to the Industrial Research Establishments in the DOI would seem a sufficient caution to research directors to consider the industrial relevance of their programs, but placing executive power in the hands of a committee of civil servants and industrialists may lead to resources being allocated more readily to projects with a clear short-term benefit over medium-term projects of a more speculative nature. If a criterion for evaluation of this policy is public accountability, then is it not wisdom to prefer the short term over the long term? It has been one of the basic axioms of government research policy that it should fund research of long-term interest to industry because it is precisely this type of activity that industry may tend to neglect. Again, it remains to be seen just what types of industrial research and development the Research Requirements Boards will call forth, but the propensity of the existing system to fund short-term research that meets a clearly perceived need of some specific subsector of industry is a hypothesis that is worth testing. If the outcome were positive, it would be entirely consistent with the strategies developed so far by the boards.

The third observation arises from the requirement that each laboratory derive at least part of its income from contract work. These contracts may be from industry directly, the research associations, or, indeed, from other ministries. Any resources so obtained lie outside the control of the boards. This requirement has led to the establishment in the IREs of marketing functions and organizations. Even though all have set up some sort of marketing unit, none has equaled the highly developed Marketing Division of NEL, but, there again, none has had quite the industrial orientation that laboratory has traditionally had. The marketing staff at NEL see their success depending closely on the active involvement of, and accompaniment by, research staff in all the stages of marketing activity. This enables the customer to receive a better service and involves the research staff with the proposal from an early stage. The latter are not, therefore, presented with a fait accompli to work on but actually participate in all stages of the formalities and execution of a project. We see here a very different application of the customer–contractor principle. In this case, a specific firm, or group of firms, is the customer, the laboratory constitutes the contractor, and some sort of price mechanism is operative.

Insofar as the empirical literature on technological innovation provides guidelines for the success of innovation, a close relationship between customer and contractor in this sense would appear to be essential.[35] On the othre hand, if close relationships between the IREs and individual firms are desirable, what is the nature of this relationship under the normal operation of the Research Requirements Boards? As we have seen, these boards should be concerned with industrial research at a higher level of aggregation than the individual firms.

Again, referring to the empirical literature, there is much evidence pointing to the importance of technology transfer in the genesis of technological innovation. New ideas are transferred from one context to another and create new possibilities for technological development.[36] The requirement boards appear to be aware of the importance of technological transfer phenomena in technological innovation but, again, one must remain skeptical about the efficiency of publicity programs that merely retail the programs of the industrial research establishments. The numerous case studies of technological innovation that have pointed out the importance of technology transfer indicate that the technology eventually was sufficiently unfamiliar initially as to arouse a good deal of resistance within the firm. The overcoming of this resistance in many respects constitutes the essence of innovative activity and takes place in an environment where the presence of research requirements boards would, in general, not be appropriate. If the results of research undertaken in the research establishments are to reach into individual firms in such a way as to actually influence decision making, then efforts of a radically different kind than the conventional form of industrial liaison will be needed.

The same problem may be approached from a slightly different perspective. If close contact between laboratories and firms is so essential in the early phases of the innovation process, why are the laboratories not required to make their living by revenue from industrial contracts? To require that the laboratories survive on the open market may seem an attractive proposition but, when one realizes that repayment income in 1973–1974 of NPL was 4% of total expenditure and of NEL, 7% of total expenditure, second thoughts may arise. The ability of a laboratory to increase these percentages substantially will depend on a number of variables, including its ability to solve industrially related problems and the extent to which individual firms see their future growth in terms of the kinds of answers that research and development can give. In any event, allocating public money into industrially related research and development can, surely, be justified only if it benefits industry as a whole. It should not, in the main be regarded as a resource on which individual firms can draw to their particular advantage. As we have seen already, this is simply another aspect of the central dilemma of the Research Requirement Board system: how to identify the research needs of industry.

The problem might be clarified by approaching it from a higher, rather than a lower, level of aggregation. There is no doubt that the Research Requirements Boards would benefit from some guidance as to the governments intentions for industry. For example, considerations such as which sectors are to grow, which to decline, and what the priorities for industrial expansion are, should be evaluated. Although these are still very early days, an industrial strategy is being discussed in the National Economic Development Council, and a statement on this topic has been issued.[37] Whether the government, industry, and the trades unions will be able ever to articulate a plan in sufficient detail to indicate general areas of research and development is doubtful in the short or medium term, but it seems all too clear that some overall picture of Britain's future industrial landscape is an essential input to the determination of research and development programs aimed at meeting the needs of industry.

13 Government Policy and Innovation In Japan

Hiroshi Inose

This chapter describes the background on government policies and practices for technological innovation in Japan. The description is of science and technology in general, and information technology, the author's specialized field, in detail. The paper also describes some of the targets for the future as well as the author's personal comments on the problems to be encountered and the method of science and technology policy for the future.

Social and Industrial Background

Japan is a small country with an area of 370,000 km. It is a heavily inhabited nation with a population of 113 million. It is a highly industrialized country with a GNP of approximately $600 billion. Because of the rugged terrain and accelerated industrialization, the population tends to concentrate in urban districts. About three-quaters of the inhabitants live in cities with a population over 30,000. No illiteracy exists and over 90% enter senior high school after finishing compulsory education, and about 30% of the high school graduates enter universities and colleges. The country essentially has no natural resources and imports almost all materials and energy from abroad.

The post-war history of Japan is characterized by the rapid growth in economy. The GNP rose at an average rate of 12% per annum until 1973 when the nation was drastically hit by the oil

Government Policy and Innovation in Japan 141

panic. Almost all aspects of industrial activities have been heavily invested in pursuit of economies of scale in production and of improvement in productivity. In the iron and steel industry, the annual output increased from 10 million tons in 1955 to over 100 million tons today. Likewise, the output of the oil refineries and petrochemical industries increased more than 10 times. The automobile industry made only 70,000 cars and trucks in 1955 and now produces more than 7 million. The output of TV and radio receivers and other electrical applicances increased by about the same order.

Technological innovation, government policy, and a motivated labor force may be cited as major factors that made possible the rapid industrial expansion. Japan adopted the strategy of quickly following up inventions originally made elsewhere and then adding significant improvements in order to establish international competence. Export-minded trade policy as well as government-oriented financial support provided opportunities for industrial build up and encouraged an export thrust. Japan's well-educated and relatively cheap labor force worked hard for reconstruction and then for a better standard of living. Another essential factor that supported the industrial expansion and the productivity improvement was the heavier than average investment in communication, transportation, and electric power.

Industrial expansion seemingly brought affluence to Japan, as intended. At the same time, however, as an obvious consequence of the rapid industrialization in a densely populated area, the environmental disruption and pollution reached such an extent that the installation of iron and steel plants, oil refineres, electric power plants, and other large industrial complexes encountered serious opposition from local communities. The industrial pattern of mass production and mass consumption in a resourceless country resulted in a trade pattern of mass importation of raw materials and mass exportation of industrial products that in turn caused conflicts with underdeveloped as well as developed parts of the world. Rapid urbanization surpassed relatively poor social investment and resulted in a shortage of housing and inferior living conditions. Burdened with affluence but with less opportunity for success, the motivation of the traditionally hard-working population was gradually lost. Thus toward the end of 25 years of rapid economic expansion, it has been increasingly felt that the industrial investment had to be balanced with the social investment and that the nation's industrial structure had to be switched from the conventional types of heavy industries that caused environmental disruption and international conflicts to something else that consumes less materials and adds more values to products.

Facts and Figures on Research and Development in Japan

To provide a background for understanding the research-and-development practices of the Japanese government, some statistics are discussed with regard to the monetary and human resources for research and development as well as the amount of international trade in technology. Corresponding figures of some other industrialized nations are also quoted with possible round-off errors for the purpose of comparison.

General Research and Development Expenditure. The total expenditure in research and development in Japan increased at an annual rate of 20% over the past 10 years and in the fiscal year of 1974 amounted to $8 billion. As depicted in Figure 13-1, Japan ranks fourth among the major nations in total expenditures. In terms of the rate of growth, however, Japan ranks at the top along with the Federal Republic of Germany. But taking into account the unusually high

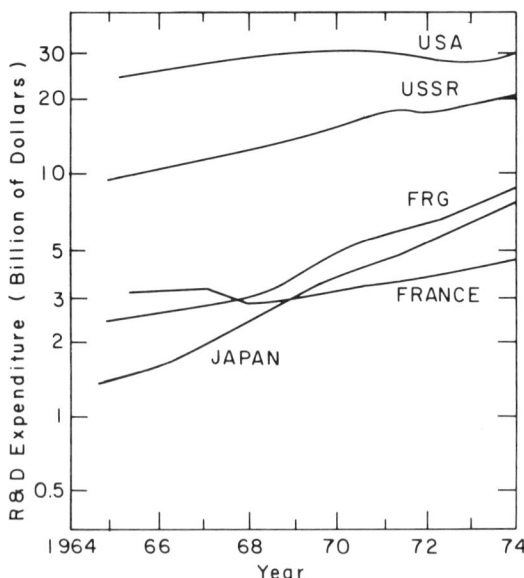

Figure 13-1. Research-and-development expenditure by year.

inflation in the past few years, the essential growth may be considered to be almost zero.

The rate of research-and-development expenditure compared to the gross national income is lower in Japan than in other countries, as depicted in Figure 13–2. The increase in the rate has slowed down in the past few years in most of the major nations due mainly to the worldwide recession.

The amount being spent in development in Japan increases year by year and in the fiscal year 1974 reached 74.3% of the total research-and-development expenditure. This places considerable pressure on fundamental and applied research. Another source of pressure is rapidly rising wages. In the fiscal year 1974, wages amounted to 50% of the total expenditure with the result of substantial dercease in investment in research-and-development facilities.

As shown in Table 13–1, Japanese industries spend 66% of the total research-and-development expenditures, whereas the universities and the research organizations spend 18 and 16%, respectively.

Research and Development Expenditure by the Government. As shown in Table 13–1, government funds mainly flow into the governmental and nonprofit research organizations and to the universities.

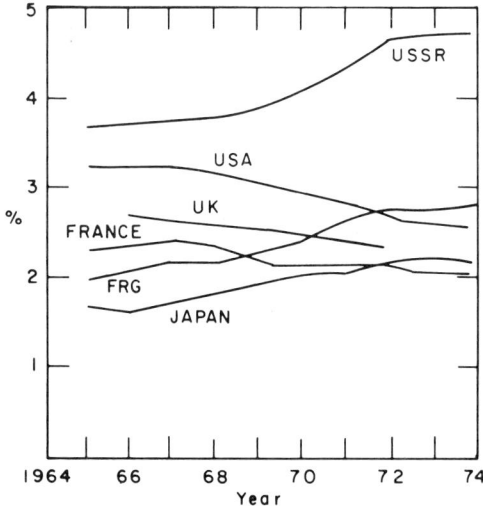

Figure 13–2. Rate of research-and-development expenditure compared to gross national income by year.

Table 13-1. Expenditures and Supplies in the Fiscal Year 1974
(in millions of dollars)

	Expenditures	Supply Sources		
		Government	Private	Overseas
Total	$8071	$2137	$5929	$5
		(26.5%)	(73.4%)	(0.1%)
Industries	$5297	$ 78	$5214	$5
	(65.6%)	(1.5%)	(98.4%)	(0.1%)
Universities	$1484	$ 967	$ 517	0
	(18.4%)	(65.2%)	(34.8%)	
Research Organizations	$1290	$1092	$ 198	0
	(16.0%)	(84.6%)	(15.4%)	

This is partly a result of the fact that the defense expenditure in Japan is very low. Table 13–1 shows that the government funds constitute only 26.5% of the total in Japan, whereas in other nations, government funds constitute about 50% of the total.

Research and Development Expenditure by Industries. Japanese industries supply 73% of the total research-and-development expenditures. The rate of industrial research-and-development expenditures to sales range from 0.2 to 3.7% as is shown in Table 13–2.

Table 13-2. Rate of Research-and-Development Expenditures to Sales

	Japan (1974)	U.S.A. (1973)	F.R.G. (1973)	France (1969)
Average	1.5	3.0	2.6	3.0
Food Industries	0.5	0.4	0.2	0.4
Chemical	2.3	3.5	—	3.4
Petroleum Refinery	0.2	0.7	—	0.9
Iron and Steel	1.0	0.5	0.6	0.4
Machinery	1.9	3.8	3.1	2.2
Electrical and Electronics	3.7	7.1	5.3	3.8
Transportation Equipment	2.1	3.5	3.1	3.2
Aerospace	0	13.5	34.4	27.6

Government Policy and Innovation In Japan   145

The average rate, which was 1.5% in the fiscal year 1974, has gradually risen from 1% 10 years ago. The rate, however, did not change appreciably in the past few years due to the recession.

Human Resources for Research and Development. The number of research and development personnel in Japan increased at an annual rate of 10% over the past 10 years and reached 240,000 in the year 1975 as is shown in Figure 13–3. The figure includes university graduates who have more than 2 years of research experience and are actively participating in research and development. In comparison with other nations, the number as well as the rate per population are considerably high as is shown in Table 13–3. On the other hand, the number of research assistants and other supporting staff are steadily decreasing in Japan. The number of supporting staff, which was 1.6 per researcher 10 years ago dropped down to 0.93 in 1975. This is less than one-half the number in France or Germany.

Because of the language and other difficulties, no appreciable brain drain has ever occurred in Japan. Likewise, no substantial inflow from the rest of the world has occurred.

International Trade of Technology. Substantial imbalance between the international import and export of technology has been observed, as depicted in Figure 13–4. In the fiscal year 1975, Japa-

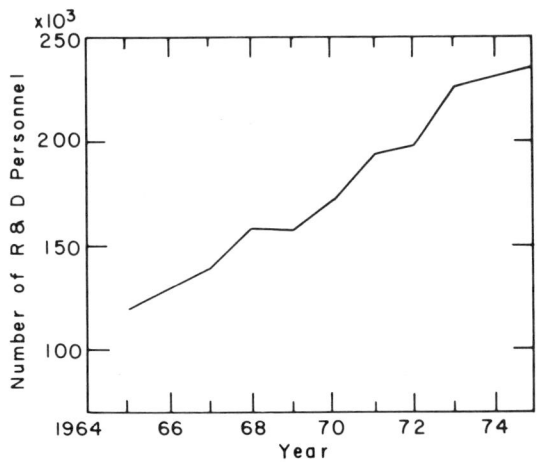

Figure 13–3. Number of research-and-development personnel by year.

146 Government: Policies and Practices for Industrial Innovation

Table 13-3. Number of Research-and-Development Personnel and Their Rate per 10,000 Population

	Number of Researchers	Researchers per 10,000 Population
Japan (1975)	238,000	22
U.S.A. (1973)	523,100	25
France (1971)	56,715	11
Germany (1974)	120,000	19
Italy (1973)	35,613	6
U.S.S.R. (1974)	1,169,700	46

nese payments for various forms of royalities was $712 million in contrast to an income of $161 million. It may be noted, however, that the payments have remained almost constant for the past few years, whereas income has increased rather rapidly. The Japanese trade deficit in technology is the highest among industrialized nations.

Government Practice

In the fiscal year 1975, the Japanese government spent $2268 million for research and development. As shown in Table 13-4 the amount

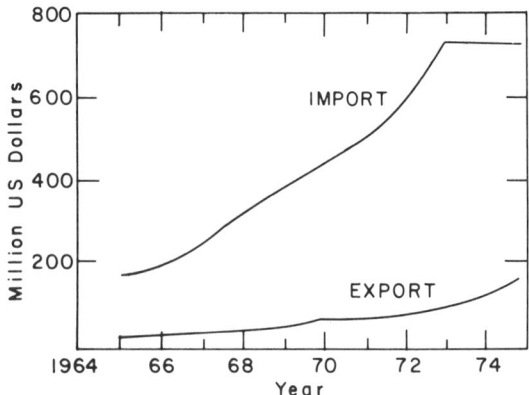

Figure 13-4. Technology import and export by year.

Table 13-4. Budget for Science and Technology

	Budget for science and technology ($ millions)	Rate to the total budget
U.S.A.	19,317	6.0
F.R.G.	4,574	3.9
France	3,710	6.0
U.K.	2,655	4.2
Japan	2,286	3.2

as well as its rate to the total budget are lower compared to other industrialized nations. As has been described, the Japanese government supports only 26.5% of the nation's total research and development expenditures. The major break-down of the amount spent by the Japanese government for research and development in the fiscal year 1975 is as follows. The Ministry of Education, Science, and Culture spent $1178 million, the Science and Technology Agency spent $566 million, and the Ministry of International Trade and Industry (MITI) spent $202 million. The expenditure of the Ministry of Education, Science, and Culture was mainly used to support 400 universities, 500 colleges, and other academic institutions, and $56 million was spent for the Grant-in-Aide for Scientific Research, which is something like NSF grants. The Science and Technology Agency spent $285 million for nuclear energy development and $209 million for aerospace development. The Ministry of International Trade and Industry spent $52 million for the National Research and Development Projects and $19 million for supporting miscellaneous research-and-development activities.

Among the government-sponsored research-and-development projects in Japan, the most outstanding are those sponsored by MITI. The MITI's National Research and Development Program, which started in 1966, has been financed by the treasury and carried out by the cooperation of government research laboratories, industries, and academic institutions. The projects under the program cover the broad area of technology for productivity improvement, environmental protection, and resource utilization. Annual expenditure for the program in the past few years has been about $50 million. The following is the list of the projects under the program:

1. Magnetro Hydrodynamic (MHD) Electric Power Generation
2. Seawater Desalting and By-products Recovery
3. Deep Sea Remote Control Drilling Equipment
4. Electric Automobile
5. Pattern Information Processing System
6. Jet Engine for Aircrafts
7. Direct Reduction Steelmaking using High-Temperature Reducing Gas
8. Comprehensive Automobile Control System Technology
9. New Method of Producing Olefin, Etc.
10. High Performance Computers
11. Desulphurization Techniques
12. Resource Recovery and Utilization Technology
13. Integrated Machining System Utilizing High Performance Laser Technology

In addition to the National Research and Development Program, MITI sponsors other projects typically represented by the "Sunshine" project, which is aimed at the utilization of new energy resources, and spent $12 million in the fiscal year 1975. The Ministry of International Trade and Industry also supports industrial research-and-development activities, a typical example of which is the new computer development project being carried out by six Japanese computer manufacturers in three groups. Another example is the very large-scale integration (VLSI) project jointly supported by the Nippon Telegraph and Telephone Public Corporation and carried out by local computer manufacturers in two groups.

In addition to financial support, the Japanese government provides such conveniences to indusrties as tax exemptions for extensive research-and-development investment and for income caused by technology export, and low interest loans for research-and-development investment.

Research and Development Information Systems—A Case Study of Government Policy and Practice in Japan

Instead of generally covering the research-and-development policy of the Japanese government over the broad areas of science and technology, a case study is made in this section in the area of communications and information processing.

Government Policy and Innovation In Japan   149

Background. Telecommunication services have grown very rapidly in Japan since the end of the last world war. Investment in the last few years has been $4 billion per annum and as the result the number of telephone subscribers rapidly increased, as shown in Figure 13-5 and reached $35 million in the year 1976. Japan is now the second largest owner of telephones in the world. The rapid growth may be partly attributed to the Japanese preference for telephones as the means of communications. The average Japanese makes four times as many telephone calls as he writes letters, the highest ratio among the industralized nations.

Domestic telecommunications services in Japan are exclusively provided by the Nippon Telegraph and Telephone Public Corporation (NTT). The NTT has been very active in research and development. The Electrical Communication Laboratories of NTT maintain some 3000 research personnel in three major locations and cover research and development in materials, devices, hardware, and software. One of the largest and the most interesting research-and-development efforts made by NTT to meet the exploding demand for telecommunications is the development of electronic telephone switching systems D-10 and D-20. Both of these stored program-controlled switching systems were developed by the cooperation of NTT and four major telecommunications manufactures—Nippon Electric, Fujitsu, Hitachi, and Oki—over the period of 10 years and manufactured under exactly the same specifications. The microwave and cable transmission systems of NTT have been developed and manu-

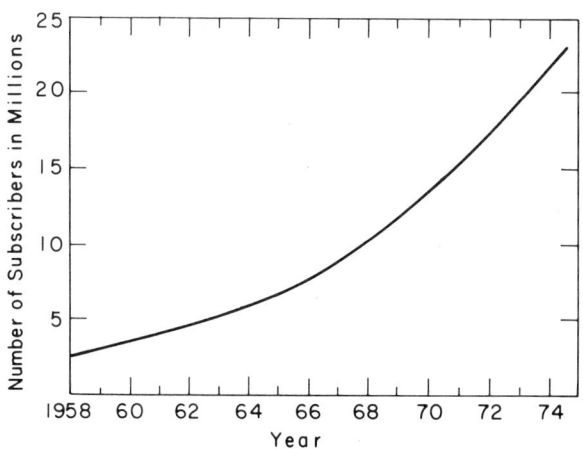

Figure 13-5. Growth of number of telephone subscribers by year.

150 Government: Policies and Practices for Industrial Innovation

factured more or less the same way by the joint efforts of NTT, Nippon Electric, and Fujitsu. This is in contrast to other nations in which the manufactures develop and supply systems of their own designs.

The computer installation in Japan has also been made quite rapidly. Figure 13-6 shows the growth by number and by value for the period of 10 years. Like telephones, Japan is now the second largest computer owner and the second largest computer-manufacturing country in the world. The Japanese computer manufacturers share 50% of the Japanese market. The Japanese manufactures have larger shares if the comparison is made in terms of the number of computers, since the computers of foreign manufacture are generally large.

In promoting Japanese computer production, the government and government-owned public agencies played an important role. The most influential has been MITI in its trade policy, research-and-development support, and reorganization efforts of computer manufactures. The gradual liberalization policy by which free trade and investment have been introduced step-by-step until 1975 provided the Japanese manufacturers the opportunities to strengthen their position. A series of contracts paid off a substantial part of the research-and-development investment made by the manufacturers. The efforts in reorganizing the computer manufacturers into three

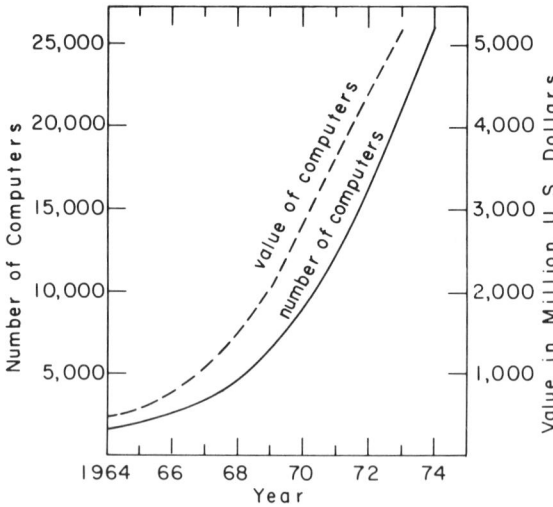

Figure 13-6. Number and value of computer installations by year.

Government Policy and Innovation In Japan 151

groups, namely, Fujitsu and Hitachi, Nippon Electric and Toshiba, and Mitsubishi and Oki, helped to eliminate duplication in research-and development efforts. The NTT has also been influential by promoting and coordinating the data communications services and by sponsoring the development of large-scale, on-line computers. Other government agencies including JNR, the Ministries of Finance, Post and Telecommunications, Education, Welfare, and others also provided wide ranges of support.

Enthusiasm from a variety of users in their use of computers may be cited as one of the major factors for computer market expansion.

Accelerated use of telecommunications and of computers naturally leads to their combined use. In addition to user enthusiasm, the fact that NTT has taken positive attitudes in providing data communication services and that all the Japanese computer manufacturers are communication equipment manufacturers as well, have been in favor of the situation. Figure 13-7 shows the growth of on-line systems by year, an average annual rate of 60%.

For data transmission purposes, NTT provides the leased circuit service, the specied circuit service, and the public network service. The second and the third services were made possible as the result of amendments to the Public Communication Law that took place in 1971 and 1972, respectively. The specified circuits are physically the same as the leased circuits but are different in that the connection of

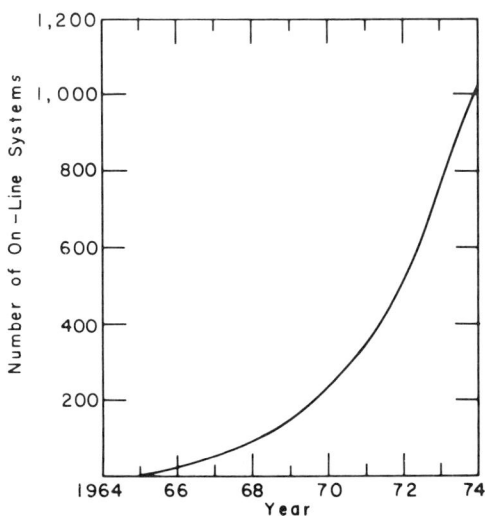

Figure 13-7. Number of on-line systems by year.

user-owned computers is possible and that the circuits can be shared by a plurality of users. The public network service is the use of NTT's switched telephone and telex networks by means of user-owned computers and terminals. To provide this service, charges for local calls, wihch used to be in flat rate, were changed to be timed every 3 minutes.

The amendment of the Public Communication Law also gave a formal permission to NTT to provide on-line computer services to the public. The services, legally called the Data Communications Facility Services, are divided into two categories, namely the public data communication services and the specific data communication service. The public data communication services are to provide the public users with the benefit of sharing large computers at a relatively low cost. The services consist of the Calculation Service by Telephone (DIALS), the Scientific and Engineering Calculation Service (DEMOS and DEMOS-E), and the Sales and Inventory Management Service (DRESS).

The specific data communication services are for government agencies and nationwide business users. By March 1975, 26 systems were in operation. One of them is the Motor Vehicle Registration Service for the Ministry of Transportation, inagurated in 1970. It connects on-line 68 officies scattered all over the country for the purpose of automobile registration and inspection. Another is the Inter-Bank Exchange Dealings System for the Federation of Bankers Association of Japan. The system inaugurated service in 1973 and connects 7200 branches belonging to 88 nationwide banks, local banks, credit banks, or others.

To facilitate the rapid expansion of data communications services, NTT started in 1968 a large-scale computer Dendenkosha Information Processing System (DIPS) Project. One of the unique features of this system is that it has been constructed individually by three major computer manufacturers—Fujitsu, Hitachi, and Nippon Electric—under the same software standard. To meet the diversified demand for data communications, NTT has been developing a new digital data network called DDX II in collaboration with four manufacturers. The system is to provide circuit-switching as well as packet-switching of data traffic up to 48 kb/sec. NTT also sponsors the development of a computer network architecture named DCNA coordinating the efforts of NTT, Fujitsu, Hitachi, and Nippon Electric.

Projects for the Future and Systems Under Development. In view of the urgent needs for changing the traditional industrial pattern

Government Policy and Innovation In Japan 153

and for filling up the increasing gap between industrial and social investment, and in view of the successful use of communications and computer technology over a quarter of a century, it has been felt that the maximum possible use of information technology in social and industrial activiites may provide a breakthrough to the problems prevailing in Japan. It has also been expected that if this could successfully be accomplished, the nation's industrial and social structure could be centered around information technology so that this heavily-populated and resourceless country could have a basis for survival and hopefully for upgrading living conditions in the coming century. In this context, the government ministries including MITI and the public agencies including NTT have been trying to set up a master plan toward an information-oriented society.

The Council for Industrial Structure of MITI, in its interim report issued in September 1974, stressed the importance of promoting the use of information technology not only in industries but also in various aspects of social activities and called for an all-out effort to establish a basis for the realization of "a welfare society with vitality." The report analyzed the needs for information technology in nine major categories of utilization, namely, government administration, industries, medical service, transportation, environmental protection, disaster and crime protection, education, distribution, and community life. As the result of the analysis, the report proposed a number of future projects including data bank and a nationwide network of government administration, computer-aided manufacture, sales and inventory system, information-sharing system for managerial planning, research and development for industries, medical information systems for community medicine, hospital automation and training of medical and paramedical personnel, computerized traffic control, computer-controlled new transportation systems, wide-area surveillance systems and data banks for environment and disaster protection, alarm systems and data banks for crime protection, computer-aided and managed instruction systems, data banks for material and product distribution, and computerized community information systems. The report also recommended such government actions as sponsoring the development of prototype models for feasibility demonstration and assessment, financially supporting and coordinate the research and development effort in the industries, encouraging the future-oriented technological innovation in research institutions, minimizing the negative social impacts through technological and legal countermeasures, and promoting the development and use of data banks and computer networks.

Other government ministries, including the Ministries of Post and Telecommunications, Transportation, Welfare, Education, and Agriculture, also have information-oriented projects by themselves or in collaboration with other ministries or agencies, such as MITI and NTT. In addition to these, the public agencies including NTT have their own guidelines for future projects.

Most of these projects are just being conceptualized or under preliminary planning. Some of them, however, are already under development. MITI, in adition to their financial support to the industries for the developing of future computers, has been sponsoring a series of national research-and-development projects as outlined in the preceding section. Information-oriented projects among them are the Large-Scale Computer Project, the Pattern Information System Project, and the Integrated Road Traffic Control Project. The first one ended up with a prototype computer, the technology of which was used in the development of commercial computers HITAC Series 8700 and 8800 and of NTT's data communications processor DIPS-1. The second one is for computer recognition of visual and audio patterns and the third one is for computerized Vehicle System Project by which an experimental personalized transportation system under computer control has been developed and was demonstrated in the Ocean Expo '75 in Okinawa.

The Ministry of International Trade and Industry and the Ministry of Post and Telecommunications jointly support the Community Cable Information System Project by which an experimental system has been built in Higashi-Ikoma Model Town near Nara and another experimental system is under development in Tama New Town near Tokyo. The objective of the project is the feasibility demonstration and evaluation of a computerized community information service by means of broad-band coaxial cables to the individual subscribers. The services include rebroadcasting of TV and radio programs; broadcasting of community TV, radio, and facsimile programs; reservation; medical information; utilities' accounting; and burglar and fire alarm services. The Ministry of International Trade and Industry also collaborates with the Ministry of Welfare in supporting the Medical Information System Project by which experimental systems for emergency medical information, rural community medical information, automated multiphase health screening, and medical data bank systems are under development.

The Ministry of Education, Science and Culture, which has been supporting the implementation of large-scale computer centers in educational institutions, is now sponsoring the development of inter-university computer network in which seven larger computer centers

Government Policy and Innovation In Japan 155

are to be connected by the NTT's new data network at the data rate of 48 kB/sec. Smaller computer centers, RJE and TSS terminals, are to be connected to the larger computer centers by means of voice-grade lines. Data bases for scientific information are also under development and in connection with these projects, a proposal was made in which a master plan for the nationwide scientific information system include general purpose and special purpose data bases, computer centers, special purpose processors, and peripherals connected by the public data communication network. The Science and Technology Agency also has a similar project, the National Information System for Science and Technology. The Ministry of Post and Telecommunications has been sponsoring the Government Administration Information Network Project, which provides, among other things, the facsimile service to the central and local government agencies.

Ministry of Transportation has the Air Traffic Control Project and in cooperation with NTT is developing a nationwide system that integrates flight and digitalized radar data processing for air traffic control and surveillance. The National Police Agency has been sponsoring the computer-controlled traffic signal systems for a number of cities. The largest of all is the Metropolitan Tokyo Traffic Control and Surveillance System, which when completed will control 8000 signalized intersections with a hierarchical computer system including 20 control computers. The Meterological Observatory is developing the Meteorological Observation System and the Meteorological Satellite System in collaboration with NTT and the National Space Development Agency, respectively. Nippon Telegraph and Telephone also collaborates with the Environmental Protection Agency for the development of Environmental Pollution Information System and with the Ministry of Agriculture for the development of Agricultural Information Distribution System.

Targets, Problems, and Approaches for the Future

Being a small but heavily populated country with almost no natural resources, Japan's industrialization and export trade has been an absolute necessity for survival. Rapid industralization, however, brought about environmental disruption and pollution. Rapid increase of importation of materials and of exportation of industrial products caused serious international conflicts. And rapid urbanization resulted in miserable deterioration of living conditions. Necessity for change in Japan, therefore, is more than any other country in the world. In this context, research-and-development efforts in Japan are now directed toward such targets as the efficient use and exploration of foods, energy, and industrial resources; the productivity improvement and value addition in manufacturing; the promotion of information-oriented industries; and the improvement in urban life and medical care. Successful realization of these targets, however, heavily depends on such factors as the available monetary and human resources and the social and technological environment. In this section, some of the targets for innovations are outlined first, and then the author's comments are described on the problems to be encountered and on the method of science and technology policy in general.

Problems. The future projects mentioned here all seem attractive provided that the social and technological environments do not undergo drastic changes, that cost and manpower for development, implementation, and operation of the proposed systems can be afforded, and that negative social impacts of these systems do not develop. Unfortunately, however, none of these provisions is realistic enough.

As the systems tend to become larger and more complex, the period required for their completion and depreciation tends to be longer. On the other hand, it is almost certain that accelerated and unpredictable changes in social and technological environment will occur in the present age of discontinuity and uncertainty. In other words, the systems face the threat of suddenly becoming obsolete from the beginning of development until after the investment is sufficiently paid off. We can depend to some extent upon the conventional methods of forecasting either by extrapolation or through consensus. But the extrapolation approach cannot cope with discontinuous changes, and the consensus approach generally provides nothing but a qualitative expression of common sense. We, therefore, have to find a novel method of modular system synthesis by which a system or its component can adaptively transfigure at any time in response to the changes in its social and technological environment. We also have to reorient the concept of compatibility toward the future and away from the past as it used to be so that the implementation of the present system does not totally jeopardize the introduction of future systems.

Larger and more complex systems require more research and development effort in terms of money and manpower. In such circumstances as the slower GNP growth and larger social investment,

Government Policy and Innovation In Japan 157

however, no substantial increase may be expected in a research-and-development budget that already exceeds 1.5% of GNP. In the country in which more than 0.2% of the total population has already been mobilized for research-and-development activities and whose monoracial population structure prevents inflow of talents from abroad, it is also quite unlikely that the human resources capable of research-and-development activities could be increased substantially. In view of these obvious limitations in research-and-development resources, it is highly important to optimally allocate the limited resources by avoiding duplicated research-and-development effort, by critically evaluating the projects to be developed, and by promoting international collaboration in research-and-development activities. Equally important is the promotion of continuing education by which the knowledge of limited numbers of talented people is refreshed so that they can actively participate in research-and-development activities.

Larger and more complex systems also require more money for their implementation and operation. Here again a reasonable policy of optimally allocating finite monetary resources is urgently required. In some of the future systems, the users may be able to pay most or part of the implementation and operation cost so that with little or some financial support, the systems could survive and hopefully could upgrade and expand by themselves. In other future projects directed toward the social welfare, however, a system could hardly be considered self-supporting or self-proliferating. The beneficiaries may be willing to accept the benefit but not always willing to pay for it, so that taxpayers should make sure whether they can afford to pay not only for the implementation of the system but also for its continuing operation, improvement, and expansion.

As for the negative social influence, the technology assessment approach extracts to a certain extent a number of negative impacts and provides possible countermeasures in terms of technological and legal action options. Unfortunately, however, the action options for reducing the negative impacts generally reduce and in many cases totally jeopardize the positive impacts the original system has had. What is important is to extract the negative impacts, to find out action options, and to show the public the result of final impact analysis so that the people are provided with the opportunities to make a choice and, through the procedure, to reorganize their own ethical standard. Such procedures for public acceptance seem to be increasingly necessary since most of the negative impacts are psychological and intended or unintended overcommitment tends to cause serious distrust in the part of the public.

Future systems, as they pursue efficiency, will accelerate the automation in industry, business, medical care, and education. Automation is good in the sense that machines generally work faster and more accurately than men, workers are released from monotonous labor, and administrators are exempted from labor troubles. But at the same time, automation may provide some serious problems. It turns a worker from producing something to merely watching the working machines. He may be bored if everything is normal. He may be frustrated if something goes wrong since the wildly running machine is too complex to be properly controlled. If many in society are employed but not actively participating in social activities they may lose their feelings of accomplishment and become unhappy. Automation replaces traditional man-to-man interfaces with incomplete man-to-machine interfaces and thereby reduces the kinship of the people and causes stress and frustration. Perhaps there should be a conservation law regarding the feeling of accomplishment. Perhaps society must provide something else to make up for the loss of such feelings due to automation. Computer-aided crafts manufacture in which each individual can interactively materialize their cerativity may be one of the possible solutions. After all, automation should be directed from productivity improvement to the happiness of the human mind.

When a system is first introduced for service, the public generally takes its benefit as a bounty. As the system continues to survive in the society and becomes larger in scale, however, social and economic activities develop and transfigure on the premise that the public deserves to receive the benefit. At this stage, a system failure causes uncontrollable chaos to the social and economic activities that now depend heavily on the availability of the system. People begin to require social responsibility from the system or even ask for compensation that may shatter the existence of the system itself. To guarantee the required social responsibility, a system generally has to have such provisions as failure detection and isolation, parallel or standby redundancy, automatic diagnosis, fall-back capability, and the like. Additional investment for these provisions, however, tends to upset the economy of the system. It would be advisable to integrate individual systems together so that systems can substitute or complement each other in case of failure. In this respect, technological standardization, cooperation across organizational boundaries, and optimum relocation of activities may be required.

14 Government Policy and Innovation in the United States

Leonard L. Lederman

This paper briefly summarizes the results of research and analysis on government policies and practices with references to the more definitive literature. It is designed to present the results of such work on the situation in the United States.

The title requires clarification in two respects. First, the term "innovation" or "technological innovation" comprises all aspects of the processes of innovation, from conception or generation of an idea to its widespread use by society. This includes all activities involved in the creation, research, development, and diffusion of new and improved products, processes and services for private and public use. In this sense research and development is most frequently a necessary—but not sufficient—condition for innovation.[1]

Second, the assignment of a title that puts "Government Policy" first and "Innovation" second should not be interpreted as a conclusion that government policies are of major significance. It is not at all clear that, in the United States, government policy is as important as many, especially government bureaucrats, think.

Finally, it should be stated at the outset that the assessment presented here is that of the author and should not be interprèted as the views of the government of the United States.

Levels of Discussion of the Role of the Private and Public Sector in Innovation

A discussion of the role of the private and public sector in innovation in the United States should distinguish between three types of cases:

1. The support for the institutional capability; knowledge base; and scientific, engineering, and technical education is generally accepted in the United States as both a public and private responsibility. At this level there is general agreement that government has a clear role especially with regard to education, educational institutions, and basic research.
2. In cases where the public sector is the eventual purchaser or funder of the output (e.g., national security, space programs), there is general agreement that the public sector has the major planning and funding responsibility but relies heavily on the private sector as the major performer.
3. In cases where the private sector (e.g., firms, consumers) purchases the outputs of innovation, there is agreement on the clear responsibility of the private sector and much dispute about the role, if any, of the government. This is a major issue for current and future debate and resolution and, therefore, will be the focus of much that follows.

Possible Rationale for Government Involvement

There is persuasive empirical evidence, although it is surrounded by significant limitations, that research and development and technological innovation have had a significant positive effect on the economic growth and productivity increases in the United States. A National Science Foundation review by a number of leading economists who have conducted research on this subject concluded that:

> Although what we know about the relationship between R&D and economic growth/productivity is limited, all available evidence indicates that R&D is an important contributor to economic growth and productivity. Research to date seeking to measure this relationship (at the level of the firm, the industry, and the whole economy) points in a single direction—the contribution of R&D to economic growth/productivity is positive, significant and high.[2]

The individual studies of the economic payoff from investment in research and development in the United States vary in their methodology and focus. Studies have been done at the level of individual innovations, individual firms, whole industries, and the national economy. Correlations have been made between research and development and productivity, and average and marginal rates of return have been calculated. For individual inventions, the rates

Government Policy and Innovation in the United States 161

of return, based on conservative assumptions, range from 10 to 50%. At the industry level, results have shown the following rates of return:

Chemical Industry	30–50%
Petroleum Industry	40%
Food, Apparel, and Furniture	15%
Agriculture	35–170%
Selected Other Industries	15–55%

Turning to the United States economy as a whole, it has been estimated that average rates of return were in the range of 30–55% in 1966. The conclusion reached in this NSF review is that:

All reasonable ways of looking at the matter lead to the conclusion that the rates of return are very high as compared to usual estimates of rates of return on capital formation.

Another major conclusion of this review is that:

Based upon the evidence, good judgment would lead to the conclusion that the United States is probably underinvesting in civilian sector R&D from a purely economic growth/productivity point of view. However, nothing can be said, based upon this conclusion, as to where particular R&D investments should be made. What this judgment means is that there is good reason to expect that a well diversified incremental R&D investment will result in high payoffs similar in magnitude to those of the past.

There is general agreement on a number of reasons why the market mechanism by itself is likely to lead to an underinvestment in research and development from society's point of view. An investment in research and development involves a large element of risk, the more so as one approaches the basic research end of the spectrum. An individual firm, or industry, may not be willing to provide funds for so risky an activity. In addition, much of the benefit frequently accrues to other firms or industries or to society as a whole. Because the firm cannot appropriate many of the benefits flowing from the expenditures, it will be less likely to make such expenditures. Nevertheless, from the viewpoint of the country as a whole, such expenditures may be highly valued.

In addition to such theoretical underpinnings, termed "externalities," we now have the beginnings of empirical measurements that seem to support the theory. A pioneering study by Professor Edwin Mansfield, of the Wharton School of the University of Pennsylvania, of 17 industrial innovations revealed great variability in the rates of return to the firm, before taxes, of about 25%. Total median rates of return to society were twice as high as the private rates of return to the firm itself. Most important for public policy was the finding that a significant proportion of innovations produced very low private returns but high returns to society.[3] This supports the externality theory, and we are now in the process of independently replicatng the results and broadening the number and types of cases.

I should mention, in passing, that inadequate venture capital does not appear to be a basic problem in the United States as frequently claimed by some observers. An in-depth review of what we know about this subject does not support the frequently stated contention that the United States government must take action to increase the supply of venture capital to innovative activities.[4] This is not to say that for some organizations, especially during downturns in the business cycle, venture capital problems do not exist. Rather, it is to point out that in the United States this is not a general problem that is susceptible to government action.

Even though the research shows imperfections in the market mechanism from the social returns point of view, this does not necessarily mean that government action is warranted, efficient, or effective. In order to justify government action, it is necessary to demonstrate that private returns are insufficient to call forth adequate investment in innovative activities *and* that proposed government policies and actions are cost-effective. In fact, the record to date of action by a number of different countries including the United States with regard to civilian technological innovation is replete with many failures and few successes.[5] However, the noncapturability by the investing firm of some of the returns from innovation plus the much higher social than private returns does offer the major rationale for possible government action. What, if any, additional public policy and action would be appropriate in this area is a major issue to be resolved. The lack of a resolution of this issue at the present time is a major reason for the lack of specific implementing policies, although many have been suggested.[6]

The Effect of Government Regulation of Business on Technological Innovation

One area affecting technological innovation in which the United States government has acted, frequently for other reasons, is regula-

Government Policy and Innovation in the United States 163

tion of business. It is, therefore, useful to review what we know about the effects of such action. In doing so it is helpful to separate two kinds of regulation:

1. Economic regulation
2. Health, safety, environmental regulation

There is evidence that, on the whole, economic regulation has at a minimum a distorting, and at a maximum a negative, effect on technological innovation.[7] No consensus exists as to whether health, safety, or environmental regulation has been beneficial or detrimental to technological innovation on the whole. There are examples of both kinds of results and good reason to believe that we should not expect general conclusions because of the differing forms of specific regulations and their varying effects.[3] One can, however, suggest a few reasonable guidelines for policy:

1. The less regulation necessary to meet specific objectives, the better.
2. The carrot (e.g., incentives) as well as the stick (e.g., disincentives) is generally a more effective approach than the stick alone.
3. The need to be sensitive to growing concern that an increasing amount of industrial research and development and innovation resources (both financial and manpower) are going for defensive regulatory purposes and being diverted away from offensive economic objectives. At some point diminishing returns in terms of the resource base for pursuing both private and public objectives may set in. But, whether the United States has reached such a point, or when it might, is open to great debate.

While patents are generally thought of as a positive incentive, they are nevertheless a form of regulation. The importance of patent rights for a firm's or industry's innovative activity varies significantly from firm to firm and industry to industry. The variance in the importance of patents may be due to the rate of technological innovation and the existence of trade secrecy laws and practices.[9]

Some have argued that relaxing antitrust regulations might promote research and development and innovations. The evidence indicates that small to medium-sized firms conduct research more efficiently than large firms. Increases in firm size, beyond some intermediate size, do not appear to be especially conducive to increased research-and-development intensity. Medium-sized to large firms, however, may offer economies of scale in later phases of innovation and are better able to exploit or develop research-and-development findings.[10] The conclusion this and other research suggests is that while there may be reason to consider changes in antitrust policy, laws, and enforcement, research and development is not a significant factor for doing so.

Available evidence suggests that the U.S. should maintain its present policy of permitting technology transfer to other countries relatively free of restrictions despite pressure to the contrary. There is no evidence that on net balance, United States economic welfare, including employment, has been hurt, and much to indicate positive effects.[11]

Conclusion

Given this research and analytical base, let me conclude with my own judgment as to what I think are the major criteria or guidelines for public policy and practice regarding civilian sector technological innovation in the United States:

1. Government policy and practice regarding civilian sector research and development and technological innovation should be made more consistent with government economic and social policy.
2. Government policy and practice should be more consistent over time and avoid constant change and uncertainties that discourage private investment and activities.
3. Wherever possible, government policy and practice should reinforce and help perfect private market forces rather than substitute for them. There is a growing body of evidence that governments have a tendency to carry such activities too far or stay involved too long. This frequently leads to:
    a. Government research-and-development substituting for and discouraging, rather than complimenting or encouraging, private investment.
    b. Interference with the positive market incentives when particular developments reach the commercialization stage.
    c. Government tending to push technology, which is usually less effective than relying more on market needs to pull innovation. As pointed out earlier, technology is not synonomous with research and development, and the justification for government efforts designed to push the frontiers of knowledge does not apply to technological developments.

15 Government Action and the Innovation Process

Albert H. Rubenstein

Since 1972, members of the Program of Research on the Management of Research and Development have been designing and carrying out field studies and field experiments in the broad area of "barriers and facilitators" in the research-and-development innovation process.[1] Some of the studies focus on specific kinds of barriers and facilitators in particular sectors (e.g., transportation,[2] energy,[3] industry,[4] others pay particular attention to government-imposed or government-related barriers and facilitators, (e.g., patents'[5] procurement policies,[6] incentives[7], still others cover a wide range of barriers and facilitators as perceived by industrial people in both the United States and abroad.[8]

The brief paper contains three illustrative conceptual models that we have used in the various studies, selected references to our papers and publications in this field, and some data from one modest study in the automotive industry. Most of our studies in this field have been guided by different versions of the conceptual models illustrated in Figs. 15–1 and 15–2.

Figure 15–1 presents a general flow model of the research-and-development/innovation process, from the early research-and-development stages through several transformation stages (dissemination, diffusion, adoption, adaptation, utilization, implementation) to pre-ultimate and ultimate impacts at the economic and social level. The impacts of government action (or inaction) are included at the various nodes of the flow diagram in Fig. 15–1, where the large number of parametric variables is illustrated in the lower box.

The entities in Figure 15–1 are:

A. *The Research-and-Development Process Itself.* This consti-

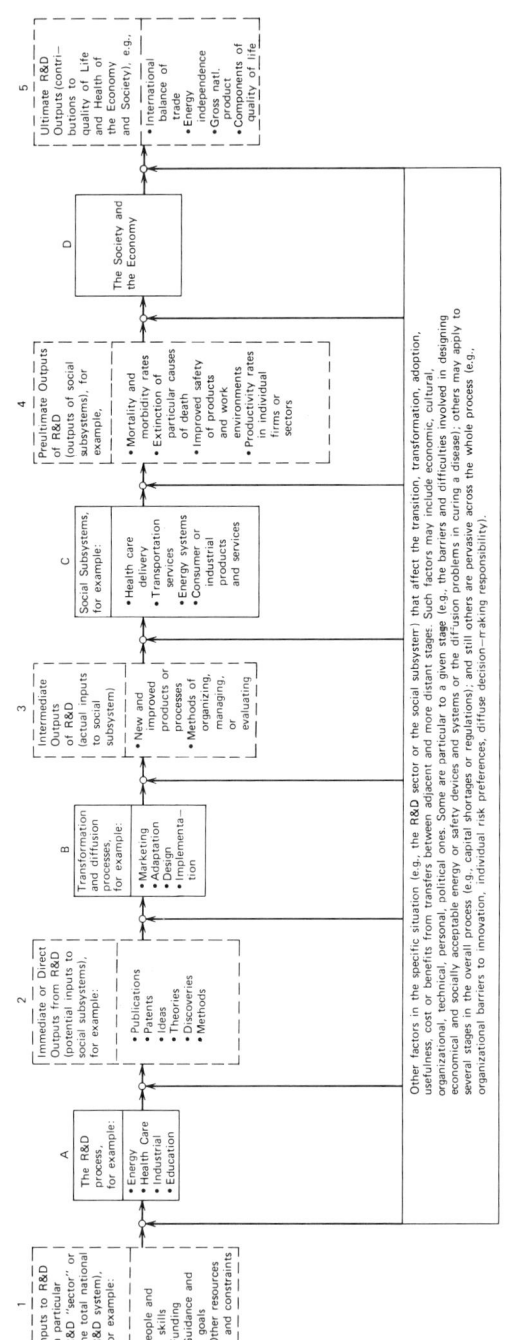

Figure 15–1. A preliminary conceptual model of the linkages between the research-and-development process and social systems.

Government Action and the Innovation Process

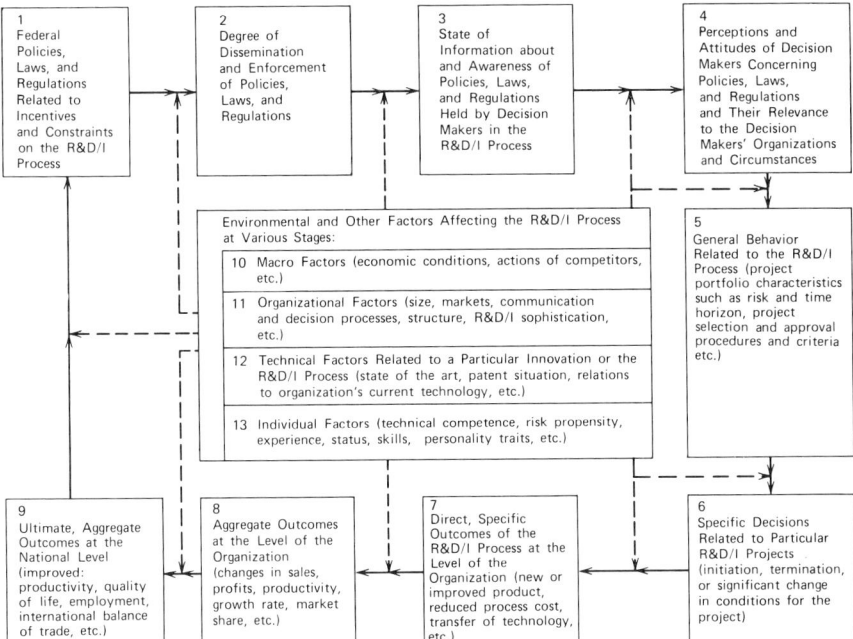

Figure 15-2. A flow model of the potential effects of federal government action and other factors on the research-and-development/innovation (RD/I) process.

tutes the social institutions, organizations, equipment, facilities, people, and procedures that make up individual research-and-development activities (e.g., laboratories); the aggregates of research-and-development activities in particular fields or sectors of the economy; or the overall aggregate of research-and-development activities in a country or other social/political unit.

B. *Transformation and Diffusion Activities.* These are, loosely, intermediary organizations and processes that are usually distinguishable from the research-and-development activities themselves.

In some cases (e.g., scientific and technical information (STI) or process design), these "intermediary" activities may be so embedded within research and development as to make them indistinguishable from their host organizations. In other cases, they are quite distinct and may be readily identified, enumerated, designed, organized, managed, stimulated, and the like, independent of the research-and-development activities with which they interface.

C. *Social Subsystems.* These are institutions and/or specific organizations or processes that are established and maintained to fill an actual or perceived social need or set of needs. Typically, they deliver products or services of one kind or another (e.g., health care, education, law enforcement, energy, transportation). They may be part of the public or private sectors or a combination of both.

D. *The Society and the Economy.* This is a shorthand way of representing the ultimate recipients or beneficiaries of the various products and services provided by the social subsystems.

These four major entities on which we are focusing may be considered linked, with respect to science and technology, by five kinds or stages of indicators that we can identify in the overall process. They are:

1. *Inputs to the Research-and-Development Process.* These constitute the resources and constraints provided to and imposed upon research and development by the environment, its host organizations and institutions, and the particular circumstances in which it operates (e.g., the nature of the field, the state of the art, the available skill levels of people in the field). The *Science Indicator* series, published by the National Science Board, has identified and quantified a number of such measures. One of the principal investigators and another member of the project team developed early drafts of the industry chapter for the 1974 volume. These input measures are based primarily on the annual National Science Foundation surveys of manpower and expenditures. They constitute the most authoritative indicators for the inputs to research and development we have, but they are lacking in at least one critical dimension—quality of the manpower and skills devoted to research and development—and one large component—the many supporting activities within the reporting organizations that are integral parts of the research-and-development/innovation process. The reasons for the latter omissions are historical and involve the difficulty of identifying and measuring

many supporting activities and the conceptual problems of what to include and what is "really not part of research and development." Some of these additional activities may be picked up under entity B in our model—Transformation and Diffusion Activities.

2. *Immediate or Direct Outputs from Research and Development.* These are the classical output indicators used to assess the productivity of the research-and-development process itself or individual activities within it. They contain, as indicated in Figure 15-1, the usual counts of patents, publications, and other indicators. In *Scence Indicators 1974*, we attempted to go beyond the traditional output measures and included such things as "innovations," "new products," and other measures that, in our model, may be one or more stages "downstream" from direct research-and-development outputs.

3. *Intermediate Outputs of Research and Development or Inputs to Social Subsystems.* These are indicators of ideas, methods, products, equipment, systems, materials, and so on, that are potentially of direct or almost direct use by specific social subsystems in their attempts to fulfill their objectives (e.g., health care delivery, education, law enforcement, or protection of the environment). This is a very fuzzy area and in some cases, particular social sectors or subsystems, there are multiple stages of transformation, adaptation, and adoption required before the research-and-development output or its consequences (e.g., a new theory of the cause of a particular disease or a new concept for an energy or transportation system) can beneficially impact the social subsystems or its own outputs. Research on diffusion in many fields (agriculture, industry, health care, social customs, education, consumer purchasing) has yielded a large number of special models and stages to explain what we have included in one aggregate entity (box B) and the linkages associated with it. Some of our research on technology transfer and related processes (e.g., information dissemination) have yielded highly elaborated models for this part of the overall process, in some cases requiring a dozen or more entities and process stages. For present purposes, however, in the exploration of indicators at various stages in the "R and D—to—ultimate outputs" process, we are representing these intermediary activities in a highly simplified manner.

4. *Preultimate Outputs of Research and Development.* These outputs (services, products, other consequences) of particular social subsystems may or may not incorporate the results of scientific or

technological inputs. As we move downstream from research and development itself (to the right in the flow model of Figure 15-1), the ability of an observer or even a participant in the relevant research and development, intermediary, and/or social subsystems to clearly identify or measure the scientific or technological component of a subsystem output becomes radically reduced. In some of our work on NASA-originated innovations, for example, we observed that neither the users, makers, distributors, nor inventors of many products or processes can clearly identify the NASA contribution and evaluate how much credit should be inputed to NASA research and development. Despite this, we are attempting to develop a methodology for detecting and, perhaps later, measuring what we call the embedded NASA technology in a product or process. This work may well yield additional indicators for this important transformation and diffusion stage in the overall multilinked process we are studying.

5. *Ultimate Outputs of Research and Development.* These are the things of value to the society in terms of contributing to its continued existence, its well being, its growth and the quality of life of its members. This set of ultimate indicators is a major focus for much of the work in social indicators currently underway.

Figure 15-2 attempts to describe the overall process by which government incentives, rules, regulations, and other actions are transformed into a basis for decision and action by industrial managers. The main argument in Figure 15-2 is that the path between the passage of legislation or adoption of regulations aimed at providing incentives for industrial technological innovation (Box 1) and any ultimate, aggregate benefit to the country from industrial innovation (Box 9) is a long, tortuous, and uncertain one, beset with many obstacles and time lags. Boxes 2, 3, and 4 represent, in brief, the continuous and complex communication process involved in informing potential innovating firms about existing regulations and policies for innovation and in affecting the attitudes and perceptions of decision makers about them and their relevance to their own organizations and circumstances. Our empirical studies in several countries have shown that even for incentive schemes acknowledged to be potentially beneficial after we described them to the respondents) the Dissemination (Box 2), Awareness (Box 3), and Perceived Relevance (Box 4) stages of the process had not occurred in a natural way, despite the long time that some of the incentive schemes had been on the government books.

For the incentive schemes or regulations (facilitators and barriers)

Government Action and the Innovation Process 171

to have an impact upon actual behavior in the firm, the kinds of behavior described in Boxes 5 and 6 have to be impacted. It is not until then that any specific direct outcomes of the government actions (Box 7) or aggregate impacts on the firm (Box 8) can be anticipated. Detection and measurement is something else.

Throughout the whole process, the environmental and other factors (Boxes 10, 11, 12, and 13) are having their impact at each stage. Indeed, our own studies, as well as the studies of others in the field of the effects of these factors, clearly indicate the dominant role they can play in the research-and-development innovation process vis-à-vis direct government actions to stimulate innovation or barriers presented by governmnet action, intentional or not.

Figure 15-3 is a simplified general model, based on selected variables and subprocesses from the more general model of Figure 15-2. It focuses on what the federal government can do realistically to affect the technological innovation or research-and-development/

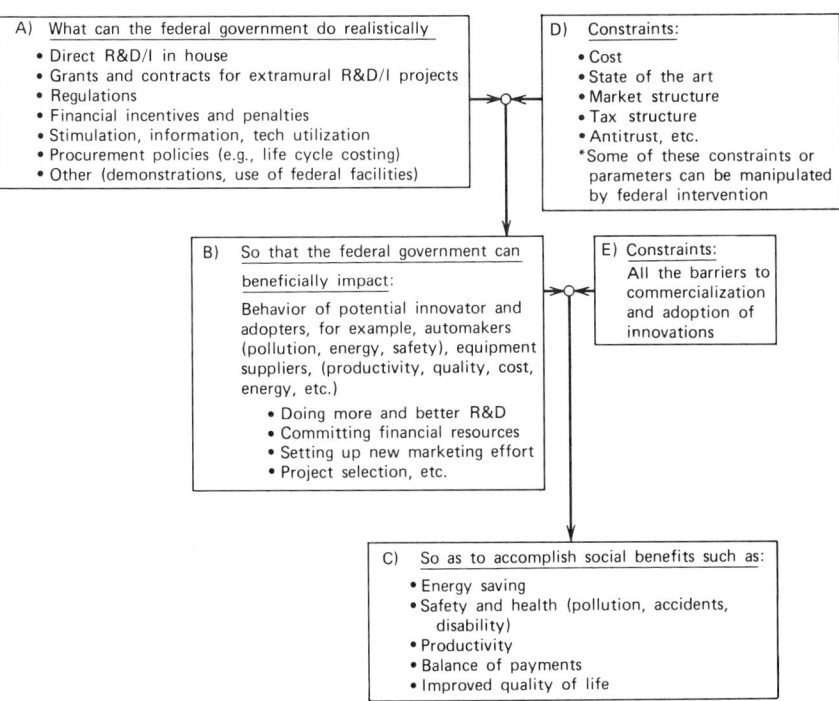

Figure 15-3. A rough, abbreviated conceptual model of the potential role of federal agencies in influencing the research-and-development/innovation process in industrial firms.

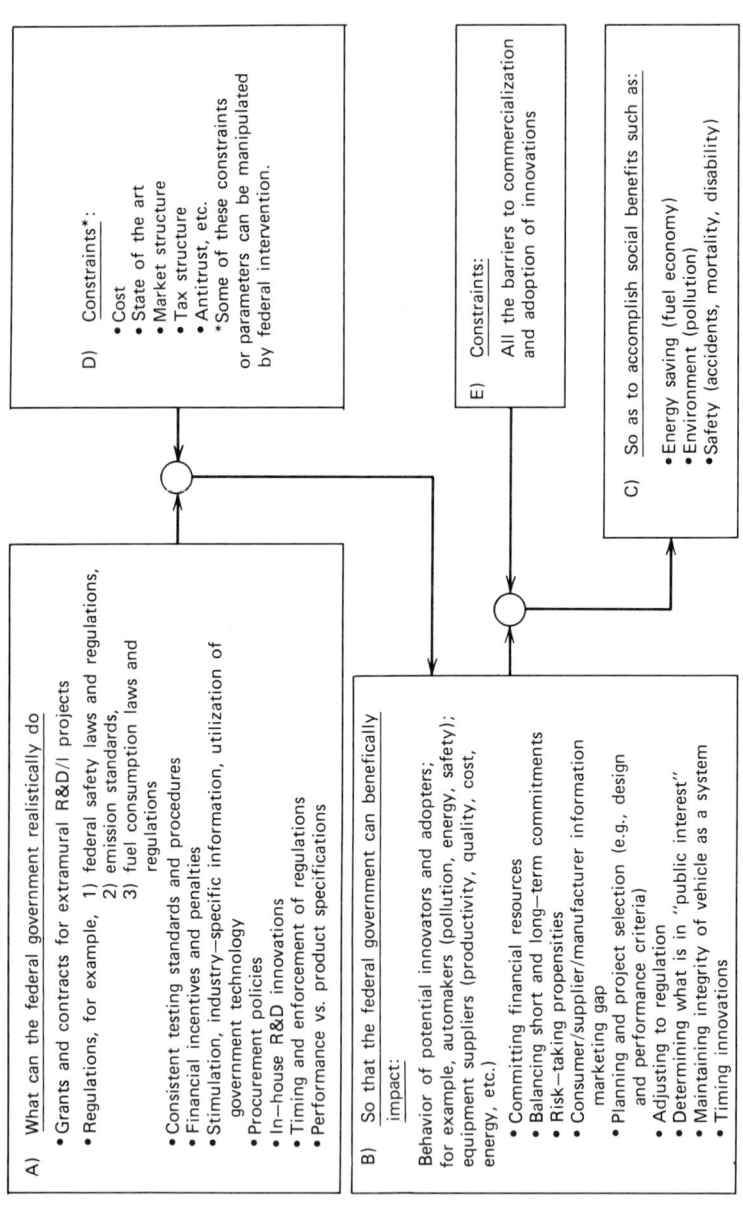

Figure 15–4. Revised model of the potential role of federal agencies in influencing the research-and-development/innovation process in the automotive industry, based on the caselette data.

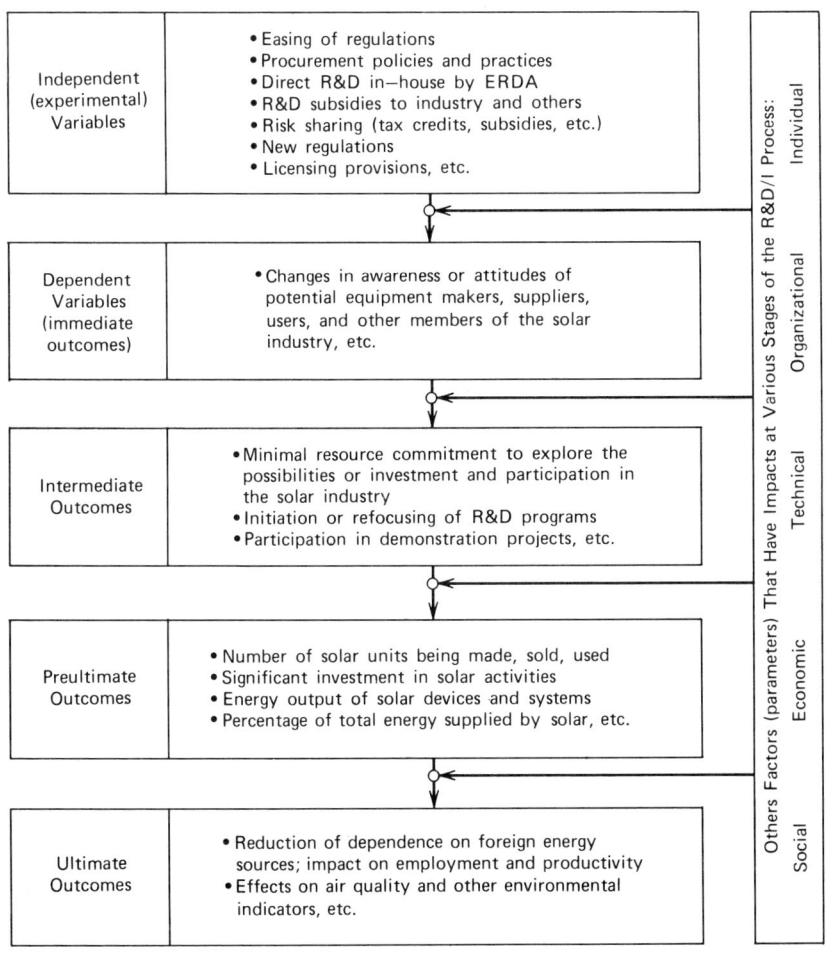

Figure 15–5. Some illustrative variables in various stages of the solar energy research-and-development/innovation process.

Table 15-1. Categorization of Facilitators

Facilitator	Number of Caselettes (Innovations) in Which Facilitator Acted	Percent (n=32)
1. Federal Law or Regulation	14	43.75
2. Challenge and Incentive of Solving a Persistent Problem	13	40.625
3. Recognition of Market Potential	10	31.125
4. Direct Government Research and Development or Grant	6	18.75
5. Technological Capability of Supplier	5	15.625
6. Federal Procurement Policy	4[a]	12.5
7. Availability of Federal Information	4[b]	12.5
8. Government Financial Incentives	3[b]	9.375

[a] Only 2 firms reported this facilitator.
[b] Only 1 firm reporting this facilitator.
Source: 77/22.

Table 15-2. Categorization of Barriers

Barrier	Number of Caselettes (Innovations) in Which Barrier Acted	Percent (n=32)
1. Federal Law or Regulation	15	46.875
2. Cost	14	43.75
3. Technical Reliability	14	43.75
4. Market Considerations	8	25.
5. Maintain Integrity of Vehicle	8	25.
6. Lack of Adequate Testing Procedure	7	21.875
7. Lack of Top Management Support	4	12.5
8. Changes in Manufacturing Process Required	3	9.375
9. Lack of Federal Interest or Competence	3	9.375

Source: 77/22.

Government Action and the Innovation Process 175

innovation process. This reduced model was used to structure a brief study of technological innovations among suppliers to the automotive industry.[9] Figure 15-4, consisting of the same structural model, contains the results of this study in terms of the factors that were found to be involved in 32 caselettes collected from 13 firms in the industry. Some of the data supporting Figure 15-4 is shown in Tables 15-1 and 15-2—facilitators and barriers, respectively.

Finally, Figure 15-5 illustrates a potential experimental design in the solar energy field, where some of the current and potential federal policies, laws, and regulations would serve as experimental variables in either contrived field experiments or natural experiments. Figure 15-7 draws upon both the family of models that started with Figure 15-2 and the rapidly expanding work we are doing in the United States and Brazil related to the concept of science or research and development indicators We currently have three projects underway or in the planning stage in Brazil based on this approach to detecting, measuring, and evaluating the outputs and impacts of the research and development/innovation process.

The models presented here involve a look into the "behavioral microdynamics" of the interaction between government and industry in the field of innovation. They have been used for research and analysis in a number of technological sectors, including: energy, transportation, health care, industry, and the environment. Such models, however, are primarily conceptual frameworks within which a great deal of time and effort are required to explicate the variables, parameters, stages, and other aspects of the research-and-development/innovation process. One of our purposes in developing and using these models is to demonstrate that linking inputs and outputs of the research-and-development/innovation process in a macro or "distant leap" fashion—from far-upstream to far-downstream variables—is a hazardous undertaking and can involve ignoring the very nature and special characteristics of this complex process.

16 Government Regulations and Innovation

Arthur Gerstenfeld

An Overview

The purpose of this chapter is to focus directly on some aspects of how United States government regulation affects the direction of innovation. In several previous studies, I considered the influence of many activities on technological innovation such as forecasting,[1] interdependence,[2] social forces,[3] and government research-and-development policy,[4] and, most recently, suggested strategies for innovation for developing countries.[5]

How regulation affects innovation is of increasing concern in the United States. It has been stated that no consensus exists on the role of regulation as a benefit or detriment to the overall rate and direction of innovation.[6] Perhaps we can shed a bit more light on that issue. This chapter attempts to separate some of the issues concerned with performance regulations and their effects on the direction of innovation.

The federal government of the United States began to regulate private activity in 1887 with the establishment of the Interstate Commerce Commission. Since that time, there has been a steady and gradual increase in government regulations that affect private industry. However, since 1970 there has been a virtual explosion in the regulatory laws which, in turn, affect technological innovations.

As shown in Table 16-1, these regulations cover a variety of circumstances. For example, the Poison Prevention Packaging Act of 1970 resulted in products such as the child-resistant aspirin cap that is discussed in more detail in a latter section of this chapter. The Clean Air Act, also of 1970, resulted in the installation of new types of filters on smoke stacks, decreasing paints and fumes emitted into the air.

Government Regulations and Innovation   177

Table 16-1. United States Government Regulations Affecting Technological Innovations (a partial list)

Year of Enactment	Name of Law	Purpose and Function
1970	Poison Prevention Packaging Act	Authorizes standards for child-resistant packaging of hazardous substances
1970	Clean Air Act Amendments	Provides for setting air quality standards
1970	Occupational Safety and Health Act	Establishes safety and health standards that must be met by employers
1971	Lead-Based Paint Elimination Act	Provides assistance in developing and administering programs to eliminate lead-based paints
1972	Consumer Product Safety Act	Establishes a commission to set safety standards for consumer products and bans products presenting undue risk of injury
1972	Federal Water Pollution Control Act	Declares an end to the discharge of pollutants into navigable waters by 1985 as a national goal
1972	Noise Pollution and Control Act	Regulates noise limits of products and transportation vehicles
1973	Emergency Petroleum Allocation Act	Establishes temporary controls over petroleum
1973	Safe Drinking Water Act	Requires EPA to set national drinking water regulations
1974	Federal Energy Administration Act	Provides authority for mandatory energy conservation programs

However, these innovations were not achieved without significant costs both in terms of direct expenses for the innovation as well as "opportunity costs" resulting from effort expended in directions that would not ordinarily have been pursued.

Having considered the increasing number of laws concerned with regulations in the United States, let us consider several examples

178 Government: Policies and Practices for Industrial Innovation

of the increasing expenses. Table 16-2 shows several examples of how regulatory expenditures have changed over time. Energy regulations went from $104 million in 1975 to $178 million in 1977—a 71% increase in 2 years. The Occupational Safety and Health Administration (OSHA) expenditures went from $196 million in 1975 to $259 million in 1977—a 32% increase in 2 years. Considering these huge expenditures—and perhaps even more importantly, considering their ripple effect on United States industry, which must spend even larger sums in order to respond to or comply with such regulation, and on the public which eventually bears the cost in terms of prices for goods and services—it is important to try to understand the implications and ultimate effects of this complicated labyrinth.

Todays Regulations—Costs

The Vice President for Research and Development at General Motors stated that 45% of General Motors' research and development was

Table 16-2. Federal Expenditures for Business Regulation
(in millions of dollars)

Program or Agency	1975 (actual)	1976 (est.)	1977 (est.)
Natural Resources, Environment, and Energy; Energy Regulation	104	172	178[a]
Commerce and Transportation—ground transportation regulation	46	52	60[b]
Health; Prevention and Control of Health Problems—consumer safety	435	493	497[c]
Health; Prevention and Control of Health Problems—occupational safety and health	196	246	259[d]

[a] Outlays for energy regulation programs will total $178 million in 1977, not including the full impact of the recently signed Energy Policy and Conservation Act.
Source: The Budget of the United States Govt. 1977, p. 85.
[b] The Budget of the U.S. Govt. 1977, p. 99.
[c] The Budget of the U.S. Govt. 1977, p. 127.
[d] The Budget of the U.S. Govt. 1977, p. 127.

Government Regulations and Innovation 179

carried out to comply with existing and proposed federal regulations.[7] Goodyear Tire and Rubber Company spent $30 million in 1974 to comply with United States government regulations.[8] Dow Chemical has estimated that the costs it incurred in 1975 to meet federal regulations was in excess of $147 million.[9] And finally, a survey by the Washington University Center for the Study of American Business predicts the costs of regulations in 1977 will be in excess of $3.5 billion.[10]

Let us now try to delve a bit deeper into the issues raised by these statements. Consider first the analysis of expenditures by General Motors as shown in Table 16-3. GM spent $1.3 billion in 1974 and the amount continues to increase. The major portion of the expenditures concern specific regulations of vehicles while the second highest expenditure is for government reports and administrative costs related to regulation.

Having considered the statements from several specific companies, let us briefly examine an industry whose innovations have been greatly affected by government regulations, namely the drug industry. Figure 16-1 shows how drug innovations in the United States have been continually declining. Part of the explanation for this poor performance is that since 1960 the costs of discovering and developing a new drug have soared 18-fold.

F. M. Scherer pointed out that "the Food and Drug Administration has gone too far,"[11] and suggests a two-tier market: government

Table 16-3. Cost Impact of Government Regulations on General Motors in the 1974 Calendar Year

	Expenditures ($ millions)	Equivalent fulltime employees
Regulation of Vehicles	884	17,500
Regulation of Plant Facilities	181	1,800
Government Reports and Administrative Costs Related to Regulation	190	4,900
Occupational Safety and Health	79	1,100
Total	1,334	25,300

Source: P. F. Chenea, "The Costs and Effects of Regulations," Research Management (March 1977).

180 Government: Policies and Practices for Industrial Innovation

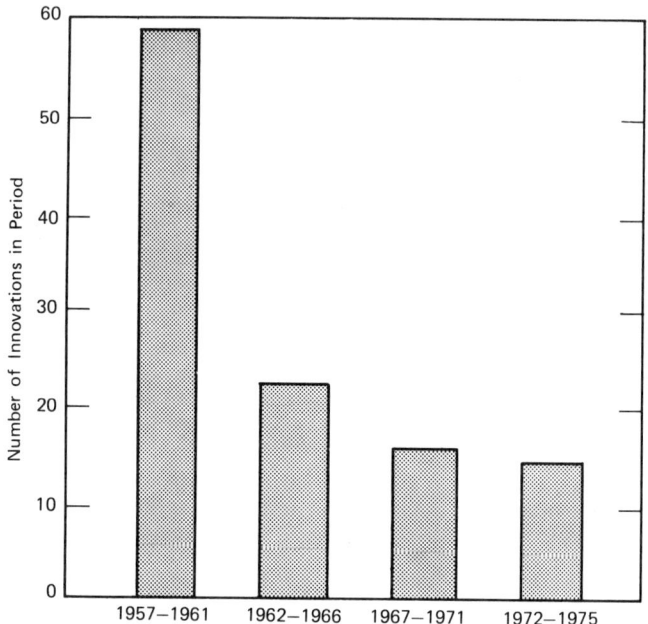

Figure 16–1. How drug innovations in the United States have declined.

approved drugs coexisting with not-yet-approved drugs. It was pointed out in *Business Week* that this would allow patients and doctors greater freedom in choosing what drug to use.

Considering the share of the drug market held by different companies by country, Table 16–4 shows that the share of American

Table 16-4. Share of Drug Market by Nationality of Company[a] (Percentage)

	1966	1970	1973	1975	Aug. 1976
American Companies	53	43	44	42	40
British Companies	23	29	30	30	29
Swiss Companies	13	14	12	13	14
German Companies	4	5	5	7	9
Others	7	9	9	8	8

[a] P. Hodgkins, "Unexpected Effects of Government Intervention," *Science and Public Policy*, Vol. 4, No. 2 (April 1977), pp. 142–148.

Government Regulations and Innovation   181

and United Kingdom companies has decreased, whereas that of Swiss and German companies has increased. As stated by Hodgkins in *Science & Public Policy*,

> In 1960, the U.S. pharmaceutical industry looked ready to dominate the world pharmaceutical industry, in the same way that the U.S. computer and passenger aircraft industries have succeeded in doing. While the U.S. industry is still a major factor 17 years later, this domination of world markets has just not happened. Instead, first the British companies, and more recently the German and Swiss, have led in growth.

My hypothesis is that the sudden and totally unreasonable stepping up of demands by the FDA, commencing in 1962, so reduced the cash flow from new products of U.S. pharmaceutical companies, and diverted so much of their R and D resources and general management time into satisfying the FDA's requirements, that they just could not keep up the momentum they would otherwise have maintained, which would have led to their domination of the world pharmaceutical industry.

Having considered the costs in terms of both direct expenditures and reduced innovations, let us now consider the other side of the coin, namely the benefits.

Today's Regulations—Benefits

Although it is difficult if not impossible to separate out the variables and establish accurate cause–effect relationships, it is, nevertheless, important to consider the effects of a regulation over time and try to establish the benefits attributable to the regulation. The traffic fatality rate in the United States has been decreasing, as shown by Figure 16-2. One might hypothesize that this is a result of the increased safety features required on automobiles. One might argue equally convincingly that the decreasing rate of fatalities is due to regulations decreasing the speed limit. In all probability both variables play a part in achieving the result, and the interaction of the two variables of decreased speed and increased safety features would be positive when the total combined effects are greater than the sum of the individual effects. By similar reasoning the interaction of the two variables would be negative when the total combined effects are less than the sum of the individual effects.

Table 16-5 shows that the United States has the lowest rate of

182 Government: Policies and Practices for Industrial Innovation

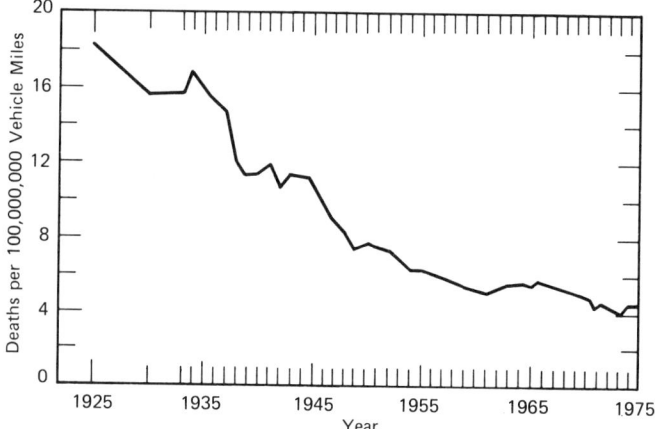

Source: J. M. Dawson, *The Love Affair with the Automobile* (Cleveland, Ohio: National City Bank, July 5, 1974).

Figure 16-2. Automobiles and Traffic Safety in the United States Traffic Fatality Rate.

traffic fatalities, France the highest ,and Canada, Italy, and Germany rank in the midle. Again, it is hard to determine the cause for such a variance, and one could espouse theories all the way from national psychological differences to equipment and road conditions. However, it seems reasonable to assmue that safety regulations in the United Statees are a major contributing factor to the decreased

Table 16-5. Selected Country Traffic Fatality Rate (deaths per 100,000,000 vehicle miles)

Country	Ratio
United States	4.7
United Kingdom	6.3
Canada	6.7
Italy	10.0
Germany	11.2
Japan	13.3
France	16.2

Source: James M. Dawson, *The Love Affair with the Automobile* (Cleveland, Ohio: National City Bank, July 5, 1974), data based on 1971.

Government Regulations and Innovation 183

fatalities, and the accrued benefits have been obviously significant. At the present time, two parallel contracts in the United States call for the development of automobile safety features still further, to design the vehicle for no injuries under head-on impact at 50 miles per hour. It remains to be seen whether this ambitious objective can be accomplished.

A particularly dramatic benefit of regulations can be attributed to the Poison Prevention Act. Accidental aspirin poisoning among young children was a major factor leading to the adoption of this law, which required manufacturers of drugs, turpentine, sulfuric acid, and other products to put safety tops on their containers. The regulations on aspirin went into effect in August 1972.

Government data since then show that aspirin poisonings and fatalities have dropped significantly. Case reports to the National Clearinghouse for Poison Control in Bethesda, Maryland, show 8146 aspirin poisonings among children in 1972, before enactment of the regulation, and 4837 in 1974, 2 years after enactment of the regulation. The cause and effect in this case appears clear as does the outweighing of benefits against costs.

The effects of regulation on innovation were studied by two researchers from M.I.T.'s Center for Policy Aletrnatives.[12] They investigated the innovations in five industries in France, Germany, Holland, the United Kingdom, and Japan. On the basis of interviews conducted in firms, the relative importance of various governmental programs in stimulating innovation was assessed. They found "that the regulations having to do with safety and environment seemed to be a positive stimulus to comercial innovations in a significant number of cases, especially in mature and concentrated industries (i.e., automobiles and chemicals)."

The executive director of the Systems Research Division at Bell Labs has pointed out that "the service-oriented regulation of telecommunications has encouraged technological innovation." Dr. Gillette goes on to point out that

> It seems that the mixture of regulation and innovation generates paradoxes. Regulatory incentives to create new products with socially desirable characteristics has a history of success and should be pursued in the future.[13]

Despite the hue and cry from the automobile manufacturers, a recent issue of M.I.T.'s *Technology Review* claimed that the accruing auto regulations are having significant effects and will have even more impact in the coming years:

By 1985 the average new American car will probably get twice the gas mileage of its predecessors of the early 1970's, will emit only about 5 percent of the carbon monoxide and hydrocarbons emitted by its predecessors of the mid-1960's, and will include safety features reducing the risk of auto travel to significantly under half the level of the mid-1960's.[14]

A Seven-Point Prescription

As technologies become increasingly complex and as society becomes more vulnerable to the effects of technology, the interaction between regulations and innovations will take on even more importance than it does today. The following brief prescription provides a series of steps that I feel should be taken in order to obtain the desired regulatory effects without inhibiting innovation:

1. It is absolutely necessary for industry and government in the United States to cooperate so that necessary regulations can work toward societal needs while the influences of these regulations on innovations are carefully examined. Industry should participate in the governmental decision making so that its input can be an important part of the government decision process.
2. During the past several years in the United States the societal impact of regulations has not been duly considered and regulations have often been imposed hastily with little regard to the ripple effects throughout society. Careful, rigorous societal impact analyses of both primary and secondary effects should be weighed prior to the enactment of each and every regulation.
3. Prior reviews are an absolute necessity, and they should be performed in a rigorous and systematic way. It should be no longer permissable to enact regulations that affect past as well as forthcoming innovations without proper examination of the full ramifications of the regulation.
4. Periodic assessment of regulations and their impact on innovations should be a standard and regular procedure. It is too easy to allow an accumulation of regulations that could have a detrimental influence on innovations. Many feel that this has already occurred in the United States pharmaceutical industry and unless periodical assessments become routine this phenomena can occur in still more industries.
5. The fact that conflicting agencies often issue conflicting regula-

tions means that present innovations are often thwarted and new ones are often discouraged. A clearinghouse system must be devised so that compatibility can be obtained.

6. Research programs and experiments should be used so that more of the uncertainty in connection with the regulation effects can be reduced. I have already recommended to the National Science Foundation a research study aimed at providing significantly more information than is presently available on the impact of regulations on innovation. There are undoubtedly large variations among industries on the effects of regulations, and it is essential that these differences be understood.

7. Since the theme of this chapter has been to consider a cost–benefit approach to regulation effects on innovations, it seems fitting that the final point should emphasize that philosophy. Even though we recognize that regulations can sometimes inhibit, sometimes stimulate, and sometimes bend innovation, we must become increasingly aware that we should not overregulate cases where the costs exceed the benefits. We must carefully evaluate both the costs and benefits of regulations and their effects on innovations both in the United States and abroad.

17 Improving the Management of Technology

Klaus Brockhoff

A recurrent result of small business research is that lack of managerial capabilities explains an overwhelmingly high percentage of business failures and bankruptcies. Even medium-sized and large companies have large problems if we consider their management of technologies or, more specifically, their management of research and development. This has been demonstrated in a number of empirical studies on cost, time, and quality deviations from original plans in research-and-development work that tend to decrease productivity on the average[1] as compared with the planned results. Such inefficiencies may constitute deadly risks for smaller companies. It is no contradiction if one points to some highly profitable and fast-growing producers of advanced technology where this does not seem to apply; in many cases, they started with a spin-off invention from some other organization that bore the risks of its development.[2]

In a profit-seeking environment, we can assume that company management is certainly not satisfied with inefficiencies in research-and-development departments, even if the departments contribute satisfactory to the company's value added by any economic standards. However, the singularity of problems that arise with certain research-and-development projects and the desire to keep the secrets of the laboratory are among the reasons that preclude companies from the systematic analyses that might lead to improvements in management. Such studies should indeed go beyond occasional case studies, singular progress reports, and discussions at meetings of research-and-development managers.

Research-and-development projects within a given discipline do exhibit enough common features to become subjects of statistical research. This research, however, has been limited to big companies that provide an adequate number of projects.[3] Secrecy is a problem in comparing companies if the questions asked touch on specifics of the management process, demanding high integrity on the part of the researcher.

I want to ask first whether additional knowledge on the management of private research and development could also be in the interest of a government. This will be discussed with reference to the German situation, which may serve to stimulate remarks on other national scenarios.[4] Second, I want to deal with a possibility of improving our present knowledge of research-and-development management by making use of the government–industry interface. This is in part a strategy to overcome the secrecy argument without, however, intending to make confidential information available to the public.

Governments spend part of their budgets to support industrial research and development. They do so to support other political and economic objectives that may be interpreted as beneficial external effects of industrial operations. As this notion is too broad to serve practical purposes, it is broken down to operative levels, where tangible policies may be formulated and controlled. I do not intend to study whether we have a theoretical background to guide such attempts[5] or to check on the consistency of such goal systems.[6]

Policy objectives of federally sponsored research and development in Germany have been expressed in a great number of official and semiofficial documents. Of particular importance is the federal government's "Report on Research" that is given to parliament every 2 years.[7] The section on "state and industry" in the 1975 report shows strategies by which the government tries to support private research and development, mostly by financing part of the research or by offering to take a financial share in the risks involved in development.[8] Let us call these direct or indirect (via tax or legal regulations) research and development policies.

Problems in the management of technology within industry are mentioned only in passing in the "Report on Research":

The principle that technical competence, managerial experience . . . of the recipients of support have to warrant the success of the project through to successful innovation, does not mean a concentration of support on large firms small and medium size firms are considered to face a multitude of difficulties in getting R and D support, among them "a lack of managerial experience."[9]

I discovered only one announcement of a remedy for this specific problem: "management help for innovation" through the "Deutsche Wagnisfinanzierungs-Gesellschaft."[10] However, even though the prospectus of this company mentions consulting with management of small and medium-sized firms or inventions and innovations, it does not go into the same details.[11] Furthermore, the company could only help those few companies that received support of the development work from this corporation.

It is only fair to mention that the federal government does more to improve the management of research and development. It supports, for instance, the RKW, a German productivity agency that consults especially with small firms.[12] However, I do not believe it would try to improve on research-and-development management through its direct or indirect research-and-development policies in any controlled manner. For the sake of brevity, let us call such an approach a meta-policy of research and development.

As a contribution to the enhancement of productivity, a meta-policy of research and development seems to be in the long-run interests of government as well as industry. Government may decrease its budget allocations to achieve comparable results. Industry may act likewise not only for its share in government sponsored projects but also with respect to their other research-and-development work. At a given level of effort, one could expect and increased flow of inventions from a better research-and-development.

Before elaborating on specifics of a meta-policy of research and development, it should be made clear that governments—more or less unknowingly, as it appears—do already act in this field, but on a level that does not meet standards that such a policy should meet. The first part of this argument may be supported by the following example.

Through document BKFT 75,[13] the federal government determines minimum requirements for cost accounting in research-and-development laboratories that are recipients of government support for individual projects. Among other things, it requires that costs of indivdual projects be accountable.[14] To insure a common and comparable procedure in calculating the cost of a project, the regulation determines how certain assets have to be evaluated and depreciated and to what degree calculatory cost or cost to cover certain risks may be taken into account. I am unaware of any study that would look at one of the following questions:

1. To what degree has this type of cost accounting led to revisions in accounting practice in the laboratories?

2. What are the behavior consequences of such a system?[15] Does it help to improve business performance?
3. Has this system changed planning and control functions in the firms?

If such studies are indeed not available, one has to conclude that government is unaware of the consequences that its regulations exert on the management of research and development, as well as on its possible influence in this field.

If one wants to use governmental support for industrial research and development in terms of support for a meta-policy of research and development, it is not enough to be aware of possible consequences and to take them as given. In addition, one has to identify fields of interest where productivity increases may originate. Furthermore, certain preconditions to identify such results are necessary. Let us turn to the second point first.

Can an experimental design be powerful enough to give us conclusions on changes in the management of technology that may be introduced at the request of a government on the occasion of supporting a project in industry. This experiment or, more specifically, social experiment may introduce a new regulation for funding projects; it may deliberately bring about changes in handling or evaluating proposals. In short, it is any change of variables intended as a stimulus to exert some specific response. Experimental design means a certain arrangement of observations that are in some way related to the funding of a research-and-development project under observation of specific constraints. The experimental design determines the quality of responses and the possibility of interpreting the results. If one wants to define and experimental design, one can either try to define minimum requirements for the interpretability of the experimental results, or one can be satisfied with some type of experiment that can be performed in practice and draws conclusions on the possible quality of the results. The latter approach seems to be appropriate here, as we can make use of three well-developed frameworks of experimental design.[16]

True experimental designs are characterized by a randomized assignment of units to experimental treatments. The support of research-and-development projects by governments is certainly not designed as a random process since it is governed by rules and regulations established specifically to avoid random decisions. Therefore, we cannot use true experimental designs.

We can also exclude pre-experimental designs from our discussion since they do not allow us to make observations that would not be jeopardized by an unacceptably large number of factors not concerned with the experiment itself.[17] Examples for such designs are after-the-fact observations only, after-the-fact comparisons between the group undergoing experimental treatment and a control group, or comparisons of one observation each before and after the experiment.

This leaves us with quasiexperimental designs. "There are many natural social settings in which the research person can introduce something like experimental design into his scheduling of data collection procedures . . . even though he lacks full control over the scheduling of experimental stimuli (the when and to whom of exposure and the ability to randomize exposures)."[18] A basic minimum requirement for data collection under quasiexperimental conditions is the rejection of hypotheses, such that advancements in knowledge may be possible. Four quasiexperimental designs are relevant for our purposes (see Table 17–1). In Table 17–1, O stands for observation, and X stands for an experimental treatment. If a group on which an experiment is administered is compared to another group, both groups are arranged in separate lines in Table 17–1, separated by a broken line. Regression discontinuity compares the intercepts of regression lines at the points of the experimental treatment, where one line is constructed from observations before that treatment and, the other from observations after its occurrence. Internal validity of an experiment refers to the possibility of recognizing any effects of the specific experiment, whereas external validity refers to the possible generalizability of the results. The reader is referred to the original literature for details. Design 3 is a combination of the preceding designs, and it is considered to be "an excellent quasi-experimental design, perhaps the best of the more feasible designs."[19] Rather than to go into details I want to stress the following points:

1. There does not seem to be hope that a feasible design of quasi-experiments could be found that would be strong enough to control for all possible sources of invalidity of results. Not even all true experimental designs can accomplish this. This is specially true with respect to generalization. In this respect, the choice of the groups subjected to observations is of particular importance.

2. Different designs control for different sources of invalidity. This may prove helpful in their selection, since not all sources of invalidity may be present in a particular situation or their influence may not be strong in a certain environment.

Table 17-1. Experimental Design for Improving the Management of Technology

Controlled (+) and Uncontrolled (−) Sources of Invalidity of Experimental Results

Quasiexperimental	Internal Invalidities*										External Invalidities†				
	1	2	3	4	5	6	7	8	9	10	11	12	13	14	15
1. Time Series O O O O X O O O O	−	+	+	?	+	+	+	+	/	−	?	?	/	/	/
2. Nonequivalent Control Group Design O X O ――― O O	+	+	+	+	?	+	+	−	/	−	?	?	/	/	/
3. Multiple Time Series O O O O X O O O O ――――――― O O O O O O O O	+	+	+	+	+	+	+	+	/	+	−	?	+	/	/
4. Regression Discontinuity	+	+	+	?	+	+	?	+	/	+	−	+	+	/	/

? = may depend; / = may depend on the specific measures used.

* 1) History (specific events occurring between successive measurements in addition to the experiment); 2) maturation (process within the respondents as a function of time); 3) testing (effects of taking a test on scores of a second test); 4) changes in measuring instruments, observers, and so on; 5) statistical regression (if groups have been selected on the basis of their extreme scores); 6) differential selection of respondents; 7) differential loss of respondents from groups; 8) Selection–maturation interaction (different rates of maturation between groups); 9) instability (unreliability of measures).

† 10) Interaction with pretest results; 11) unrepresentative responses of those that undergo the experiment; 12) reactive effects of the experimental arrangements themselves; 13) multiple treatment interference; 14) irrelevant responsiveness of measures; 15) irrelevant replicability of treatments, as some of the complex components may be missing.

Sources: Campbell, "Reforms as Experiments and Stanley, *Experimental and Quasi-experimental Design*.

3. Each design necessitates observations before and after the occurrence of an experiment. This may prove difficult to achieve in practice. It does not seem to cause legal problems if recipients of government funds are required to report extensively on matters that relate to this funding after having received a contract.[20] However, it may be debated whether comparable observations could be taken for a time before the company bid for a contract. To resolve such a question, one could search for a parallel. Tenders for construction work usually do have to prove their experience in related work while bidding on government contracts, a situation that appears similar to the one in question. However, it might be necessary to have company assertions checked by outside investigators, if only to ensure proper application of analytical techniques. Firms that enter certain contracts do so to perform a research task, not to demonstrate their skill in applying a certain management technique.

4. Designs 2 and 3 require observations in organizations that do not undergo the experimental treatment. This may cause little difficulty where a research contract is awarded to a large company, with different laboratories working in roughly the same field. If one wants to extend the external validity to small and medium-sized firms, one possible solution may be to announce that all those who bid for contracts must declare that they are willing to furnish certain data (specified in advance), whether or not they win the contract. To avoid discrimination against less profitable companies, the data collection may be done at the expense of those interested in a specific quasiexperiment.

Items 3 and 4 may serve additional purposes. They break the ties of the singularity and secrecy arguments. This makes clear why this interface is of special interest for the analyses.

A successful application of quasiexperiments makes it necessary to concentrate on a few subject areas of interest to the sponsor as well as to the recipients of research funds. The interest seems to be stronger with higher possible productivity increases resulting from the knowledge derived from an experiment.

One such area appears to be the application of certain network techniques in planning, which are not accepted as helpful by all companies. Another might be the forecasting of research-and-development project cost, given a certain time frame, project area and project specification, because:

1. We know of considerable discrepancies between planned and realized project cost, if the task description in terms of the completion date and a certain quality of the solution are upheld.
2. From earlier studies, we have assembled a body of hypotheses on potentially influential independent variables that might determine cost as a dependent variable.
3. On the basis of results referred to in 2, it is possible to make recommendations for research-and-development management that are not yet in widespread use.
4. The recommendations could be made the subject of contracting and bidding procedures, since they can be based on presently available accounting information and on fairly standard planning techniques.
5. The requirement to make use of certain recommendations from (3) could be considered a quasiexperiment to be administered on a certain number of contracts. Careful interpretation of the results could gradually improve the management of research and development.

To make these thoughts more convincing, it is worthwhile to elaborate further on items 2 and 3, hoping to add some flavor to the ideas.

Let us consider the cost of a research-and-development project, C, as P is dependent on the specific project task:

$$C = f(P) = f(p(x)), \text{ with } P = p(x)$$

It is plausible to assume:

$$dc/dP > 0; \lim_{P \to 1} C \to \infty$$

With respect to the second derivative, we find rivaling hypotheses in the literature: $d^2C/dP^2 > 0$ is the early learning hypothesis, whereas $d^2C/dP^2 < 0$ would point at late learning. The kind of learning determines the fruitfulness of parallel research strategies.[22] The problem is to know where a certain learning type may be likely to occur. One hypothesis is that early learning is typical for more basic research.[23] To collect data on a possible falsification of such a hypothesis we could make use of §13, 1, BKFT 75: quarterly or biannual reporting requirements comprise information on expenditure and time. Furthermore, the answer to the question whethre the chances of reaching the project objectives have changed since the last report could be given a specific format.

Let us now specify the project x by[24]

$$x = (\|y\|, y/\|y\|, t),$$

where $\|y\|$ is a measure of distance between the state of the art and the project objectives, both depicted in a m-dimensional space of project characteristics, the technical performance space.[25] Consequently, $y/\|y\|$ is the direction of the developmnetal movement in this space. Consider t to be development time.

The following hypotheses may lend themslves for immediate analysis: $\partial P/\partial \|y\| < 0$ and $\partial P/\partial t > 0$. However, I concentrate on the more interesting concept of the technical performance space. The following questions may be studied and answered in quasiexperiments:

1. To what degree has the concept of a multidimensional technical performance space become a feasible planning concept in the age of departmentalized research and activity networks?
2. Is it feasible to estimate cost, time, and probability of success along individual dimensions of the technical performance space?
3. Does the aggregation of such estimates lead to improved planning of the project,[26] in terms of decreased cost or time overruns?
4. Does the concept of setting developmental objectives in a technical performance space facilitate communication and harmonization of planning with the marketing department, where multidimensional scaling procedures may have led to a similar space to identify possible gaps in a market?

Obviously, such questions could be answered from results of quasiexperiments, if the idea of the technical performance space become one basis for the project definition. According to §3 BKFT 75, the working program then becomes part of the contractual agreements, changes of which have to be reported and agreed upon. A report on planning and execution of the project (§13, (3), c BKFT 75) may be used to document results.

It is in passing only that we mention again the possibility of checking on the fruitfulness of certain regulations within procedures and guidelines, such as the BKFT 75 through quasiexperiments. As procedures differ with the percentage of governmental support for projects, we have a natural starting point for multivariable analyses of the fate of the projects.

I hope that it has been possible to convince the reader of the potential fruitfulness of the idea that governments should not be satisfied with trying to enhance productivity or quality of life by direct or indirect financial support of projects. In addition, they should consider improving the management of technology by explicitly engaging in quasiexperiments and their analyses rather than using pre-experimental designs to analyze and to formulate policies. It would make better use of data collected, and it might lead to procedural changes as a consequence of the experimental design.

Practitioners tend to argue that the world is too complex and changing too fast to be subjected to quasiexpreiments. As long as the changes considered concern the independent or experimental variables, changing them is at the discretion of the government. If changes occur frequently, the question remains whether this is a result of learning or whether it is a result of a felt need for change that is only imperfectly evaluated. The rest of the argument speaks more for rather than against quasiexperiments. The more complex the environment, the more there is to be controlled and the more necessary is it to adopt a design that promises to do this. A close comparison reveals that the extra cost seems minimal. It has become a standard argument to point to the fact that John Stuart Mill in 1843 already argued that a "plurality of causes" and an "intermixture of effects" could not be analyzed by deductive reasoning in the form of simple observation (which would be the nearest equivalent to the social experiments studied here).[27] The counterargument could use the fact that simple observation is a pre-experimental design, which is inferior to quasiexperimental designs.

Unfortunately, negative results are interpreted as having caused cost without having received a benefit, thus wasting taxpayers money. It may be necessary to explain the potential gain in information from a quasiexperiment irrespective of its outcome. Another argument can hardly be refuted. Expected results from quasiexperiments may threaten presently held opinions, prevailing power structures, or decision-making procedures more than pre-experimental designs because they leave less room for counterarguments. This may cause political opposition to such designs.

18 Closer Integration of Science Policy and Economic Policy

Richard R. Nelson

With a few exceptions, thinking about science policy and economic policy tends to occur in different government offices. For many years, the Office of Science and Technology and the Council of Economic Advisors were housed in the same building, but there was little interaction between the agencies. The Office of Management and Budget (OMB) has not been particularly concerned with exploring whether the research-and-development parts of a department's proposed budget, and the non-research-and-development, are intellectually coherent. And, in general, they are not. The same separation of policy design functions that occurs in the executive offices occurs within most government departments. Within Health, Education and Welfare, the National Institutes of Health stand proudly apart from the hurly-burly about national health policy. Within the Department of Housing and Urban Devedopment (HUD) and the Department of Transportation (DOT), the research-and-development operation is removed from the main arena of action. The design of environmental regulations and the funding of environment-related research tend to be viewed as independent functions and have gone on within different agencies. Within agriculture and defense there has been some coherency between research-and-development policy and other policies, but these seem to be exceptional cases.

It is increasingly evident that the separation is a serious obstacle to the design of sensible polices. The policy adopted during the early 1970s concerning automobile emissions proceeded without

Closer Integration of Science Policy and Economic Policy   197

a well-thought-through research-and-development component. The policy was vulnerable to no action on the part of automobile companies since it provided for insufficient governmentally funded research and development even to enable checking the varacity of the claims of the automobile companies. The objective of keeping down housing costs is unlikely to be achieved without significant technical advances in housing. With the exception of sporadic efforts, like Project Breakthrough, no policy has been designed to improve the pace of technical progress in housing. The SST, for example, would have threatened environmental values and resulted in a plane that airlines would not have bought without significantly increasing airfares. After the energy crisis of 1973, the articulated research-and-development policy was that profit incentives would lead to private research and development, but the pricing policies adopted made private research and development on new energy technologies appear unprofitable. The Carter energy policy has only been articulated in its economic dimensions. His research-and-development policy, which presumably will be sketched later, may be closely tuned to the economic package, or, more likely, it will have little to do with the economic policy package and will represent a continuation of the old bad habits of thinking about resarch and development separately from the other dimensions of policy.

There are some bold words in the National Science and Technology Policy, Organization, and Priorities Act of 1976, suggesting that Congress has at least begun to be aware of the problem. According to the act, the Director of the Office of Science and Technology Policy shall "work in close consultation and cooperation with the Domestic Council, . . . the Council of Economic Advisors, the Office of Management and Budget." This directive is easier said than done. And while Congress apparently senses that there is some kind of problem, it is not clear that it knows the nature of the problem, or that among other things, it is part of that problem. The uncoordinated programs were passed by Congress and, in many cases, designed there. There is reason to believe that the words about the director of the Office of Science and Technology Policy interacting with economic policy makers carries the meaning of "having greater influence" rather than "being more responsive." But both are needed, not just the first.

In the remainder of this chapter I, first, discuss why science policy tends to stand separately from other aspects of policy. Second, I examine why this separation is especially costly given today's policy problems and those problems that are likely to be with us over the next decade. Finally, I make a few remarks about the kinds of organizational changes that seem required.

The Tradition of Independence

Analysts of science policy sometimes have pointed out that such policy involves three roughly distinguishable dimensions. First, science policy is concerned with the health and support of the scientific enterprise. Second, science policy seeks to bring high-level governmental decision making objective, scientific, and technical expertise. Third, science policy aims to guide the allocation of research and development to better achieve national objectives.

It is important to recognize that even the term "science policy" is relatively new. Prior to the end of World War II, very few governments had an articulated science policy. The early conceptualization of such a policy defined the scientific enterprise quite narrowly—mainly as research scientists in universities—and viewed support in terms of relatively unconstrained basic research funding. The effort of marshalling scientific expertise was limited to matters of weaponry and disarmament. The allocation and guidance function primarily involved a few large technological enterprises under government auspices, mostly concerned with defense or atomic energy. In addition to science policy at the level of the executive office, there were as well special science policies concerned with health and agriculture contained in the relevant departments and agencies.

Regarding the two functions of supporting science and enlisting scientific expertise, a certain separation from the day-to-day political pressures on government is quite appropriate. Long-run support of the science community, as through the Office of Naval Research, the National Institutes of Health, and the National Science Foundation, requires a certain steadiness of purpose and an apolitical stance. The appropriate institutional arrangements for the advisory function are trickier. It is important to avoid pressure and prejudice but also to assure access to the ears of those who are making policy. The separation of the prime locations of the two functions that occurred during the 1950s and early 1960s, with the former placed largely in the National Institute of Health (NIH) and the National Science Foundation (NSF), and the latter in the office of the President's Science Advisor, reflects this difference in the appropriate institutional structures.

The function of guiding research-and-development allocation is

Closer Integration of Science Policy and Economic Policy 199

something else again. To be effective, it requires the kind of connection with broader aspects of policy making that a corporate-applied research-and-development laboratory should have with the broader corporate structure. This third function of science policy involves the use of reseach and development to achieve goals that ultimately are defined in nontechnological terms such as enhancing defense capabilities, improving health, or increasing productivity. It can be differentiated from the support function in that the allocation of research and development is not left largely to the decisions made by the nongovernmental community; rather, influencing allocation in certain ways is the objective of policy. It can be differentiated from the advisory function in that the government is trying to influence what is happening in science and technology. In the remainder of this paper I use the term "research-and-development policy" to refer to this allocation function of science policy.

Research-and-development policy is concerned both with the design and management of research-and-development programs funded by the government and with molding incentives for nongovernmentally financed research and development (as through providing tax credits for certain kinds of research and development. changing aspects of the patent system, adopting performance standards in contracting for a particular kind of product, etc.). The objective of research and development may be the improvement of capabilities to perform certain missions assigned to the government (defense) or the enhancement of certain important national objectives not associated with particular governmental agencies (improving general economic productivity). The scope and nature of research-and-development policy depends on two kinds of perceptions. One is the range of national objectives the government is concerned with furthering for which the allocation of research-and-development resources is seen as an important variable. The second is the extent to which a desirable allocation of research-and-development effort is seen as coming about naturally, through private research-and-development efforts, without specific governmental action to encourage that allocation.

In the first decade after World War II, research-and-development policy was quite limited in scope. The focus was largely on research and development to further certain governmental missions and programs. The prominent ones related to the long-standing governmental mission of enhancing agricultural productivity, defense, and the development of atomic energy. In the first two areas, the departments involved had a healthy appreciation both of the role of research and development in furthering their long-range objectives and of the need for a conscious research-and-development policy to achieve these objectives. In the case of defense, this appreciation was reinforced through the Presidential Science Advisory Committee (PSAC). In the case of atomic energy, the key agency involved was concerned almost exclusively with research and development, and at that time the objective of developing peaceful uses of atomic energy was both broadly defined and widely accepted. Regarding health-related research and development, the NIH could have defined its role in terms of significantly influencing the allocation of research resources. In fact, it chose not to do this but rather to define its role in terms of support of the biomedical research enterprise, with the scientific community preserving considerable autonomy over alloaction.

Because the scope of research-and-development policy was so limited in the early days, and because coordination between research-and-development policy and other aspects of policy occurred relatively naturally in the two agencies where it was important, the issue of the need to coordinate certain aspects of science policy with economic policy was not broached early. The tradition developed of a science policy standing apart, and the separateness came to be seen as a desirable state of affairs. The debates about the need to shield the NSF and the NIH from short-run political pressures were fierce, and the victory for the scientists involved was celebrated. The PSAC was widely viewed by the science community as a place to exert high-level influence on policies, as a vehicle to monitor or lead government, not to join it.

Nor was the problem seriously raised when, in the early 1960s, various science policies began to be justified in terms of their contribution to economic progress. Although this objective was sometimes mentioned in the early 1950s, it was not emphasized. However, the National Aeronautics and Space Administration (NASA) was advertised ringingly as an agency whose programs would contribute to the general technological and economic strength of the country; to a considerable extent Project Apollo was politically sold in terms of its expected "spillover." More generally, the science policy community latched on to the economists' propositions about the key role played by technical advance in economic growth, and about the externalities created by research and development, which provided a broad-guage, new rationale for public support of research and development. However, the economists' arguments did not seem to point sharply to any specific programs nor to a need to coordinate science and economic

Closer Integration of Science Policy and Economic Policy 201

policies. Thus the early ventures of science policy into this new arena stressed the desirability of more general support of science and technology—a natural extension of arguments used earlier regarding academic science. The notion that the allocation of research and development might matter, and the policy might be used to influence that allocation, a notion that would have driven science and economic policies closer together, was there only subliminally.

However, by the mid-1960s, the idea that Project Apollo, rather than advancing general economic progress, might be pulling research-and-development resources away from uses of higher economic priority began to gain currency. The middle to late 1960s also saw the beginnings of recognition that long-run economic development problems of the United States ought to be defined sectorally, rather than aggregatively, and that an important task of science policy might be to help advance knowledge and technologies in certain key economic sectors where problems were not being met by the existing regime of research-and-development institutions, or opportunities not being seized, or both. Several aborted efforts eminated from the Department of Commerce to develop a policy toward "civilian technology." The late 1960's saw a very sharp rise in government research-and-development spending in agencies such as DOT, HUD, Environmental Protection Agency (EPA), and later the Energy Research and Development Administration (ERDA). There also was a sharp rise in concern within the science policy community about certain specific aspects of private sector economic performance, particularly the slow down of the rate of productivity growth after 1968 and the declining share of American manufacturing exports in world trade. Various commissions were established to report on what could be done to stimulate research-and-development activities to remedy these problems.

Perhaps not surprisingly, the organization for thinking about and designing science policy in these new areas tended to follow the mold set earlier. The isolation of the new, evolving science policy interests is a striking phenomenon. Research-and-development programs tended to stand apart—isolated and ineffective. There are many examples of science policy agencies viewing economic problems independently, and thus futily.

I have been discussing the problem as if the isolation of science policy were largely the result of the leanings of those concerned with science policy. That is half of the story. But the other half is the inability or unwillingness of the economists and lawyers, who dominate other aspects of the policy apparatus, to see the relevance of research-and-development policy to their concerns. The lack of a research-and-development policy associated with the Clean Air Amendments, for example, reflects lack of interest, or at least uncertainity, of the lawyers and economists at EPA about the need for such a policy. Economists' models of the energy sector include a variety of possible future technologies not yet developed. These models have been used extensively to guide thinking about what the energy future will be like and have strongly influenced the design of policies. But these models tend to pay very little attention to the mechanisms by which these new technologies might get developed. And this lack of analytic attention is carried over into lack of thought about a research-and-development policy.

The Problem Today

Subtle but profound changes have occurred over the past decade in the way thoughtful people think about economic progress and about the role of science and technology in a progressive economy. In the early 1960s, economic growth meant increases in GNP. Today much more attention is paid to the composition and quality of economic activity and to the long-run consequences of patterns of economic life. The focus is less on the marco aspects and more on key sectors of problems: energy, housing, medical care, environmental protection, urban blight. There is much broader recognition now than earlier of the importance of balancing market and nonmarket values in judgments of what is the best thing to do. There is far greater awareness that the long run may not take care of itself adequately and that a central task of government is to lengthen the time perspective operative in decision making. All these developments both increase the importance of science policy and make it imperative that science policy and other aspects of policy be much better integrated in the future than they have been in the past.

The goals that the nation has set for itself will require a strong and effective research-and-development effort for their achievement. And that research-and-development effort is not likely to evolve in the absence of well-thought-through policies to make that happen. The economists and lawyers who tend not to see technical advance as an important part of any solution, or who believe that technical advance will be more or less automatically drawn through the market, are fooling themselves. So are scientists who believe that an effectve science policy can be designed without close and effective interaction with those who are designing other aspects of the policy. The evolving energy situation is a striking example. There certainly is some leeway afforded by prevailing technologies to enable us to find substitutes for gas and oil, and to conserve on energy use generally, without dramatically rising costs. However, nearly everyone recognizes that over the long run, success in coping with the problem will depend on our ability to evolve appropriate new technologies. For a variety of reasons, this country is unwilling to let appropriate be defined exclusively by the market. Even though analysts may be divided on the extent to which regulation should proceed through taxes rather than quantitative restrictions, it is highly unlikely that the tax route will be the only one taken to protect environmental values, safety, national security objectives, and the like. Thus research and development will proceed within a regime of market incentives recognized not to fully reflect social values in any time. And regulatory constraints will evolve over time partly in response to new appreciations as to what is technologically possible and not unduly costly. Effective implementation of an energy policy requires a research-and-development policy tuned to it. The appropriate mix between private and public financing of different kinds of work is an extremely important matter to think through.

Similarly, with respect to protecting environmental values, some gains in reducing automobile and industrial emissions can be achieved at tolerable cost using existing technologies. But the degree of success we can hope to achieve and the cost of achieving what we do will depend largely on the nature of the new technologies that evolve. Tax and subsidy systems perhaps should play a larger role in the regulatory regime than they presently do. But it is highly unlikely that we shall disband the use of regulatory restrictions. The experience with automobile emission control regulation should have taught us the folly of relying fully on the market to pull the appropriate research and development. But it is hard to see even the early glimmerings of a coherent research-and-development in this area.

Almost all images of the society we are striving toward require improved housing standards and better means of intraurban transportation. It is unlikely that high goals can be met without extraordinarily high costs unless significant new technologies are developed. It is certain that these new technologies will not be developed by private enterprise alone.

In each of these cases, economic and science policies need to be designed together. Economists must understand that technologies will not be static and that influencing how they evolve is a key part of the policy task. Scientists must understand that autonomy of science policy is a false value; the problem is to guide allocation rather than merely to provide advice and support. Good policy design requires strong understanding of what kind of research and development we can expect to pull through market incentives and what kind of research and development needs to be more directly encouraged, funded, or undertaken. An effective policy must be based on sophisticated perception of the criteria used by the operating agencies—business firms, households, government agencies—in their decisions to adopt or not to adopt a technology and the ways in which, and the limits to which, incentives can be influenced. In other words, science policy must be developed in the context of sophisticated economic analysis of the situation, of the policy goals, and of the range of policy instruments that can be used to achieve them.

The Organizational Problem

The current institutional settings are not such that these understandings will develop. Nor will they be easy to foster. However, it is possible to delineate some of the actions that should be encouraged.

The re-establishment of a science policy office in the executive offices provides an opportunity, but one that unless carefully nurtured is likely not to be seized. There is a real danger that the inclination in the new office is to go back to the situation that existed in the late 1960s. But if the prospective proposed here is correct, that situation was grossly inadequate. The problem is only in small part to establish a presence for science in the executive office. The real problem is to mesh a perspective on what can evolve technologically, with a sophisticated assessment of the appropriate role of the federal government in facilitating the evolution of that technology, in the design of policies more generally. Something like an institutionalization of meetings and discussions among the Director of the Office of Science and Technology Policy (OSTP), the Chairman of the Council of Economic Advisors (CEA), and the Director of the Office of Management and Budget (or their high-level representatives) seems appropriate. The older executive office science policy machinery was designed to coordinate science policy across government agencies, not to coordinate science policies with other policies. The latter is the great need today.

A change in perceptions in all three of the key agencies is required. The Council of Economic Advisors has evolved over the years into an organization much more concerned with microeconomic policies than it used to be. But the economics profession tends to be blind to the importance of technical change at the microeconomic level and, hence, to the importance of considering that variable very seriously in the design of microeconomic policy. For the new OSTP, the notion of seeing a science policy is one instrument that needs to be tailored to fit an overall policy design will be both hard and painful to grasp. It implies a considerable loss of autonomy. It is the OMB, of course, that is in the best position to require that the design of an overall policy specifically involve a research-and-development component, and that science policy be linked into an overall policy design. This is not the way OMB is used to thinking, however.

Similar organizational problems exist at the departmental level. In some departments, the science policy component has established a strong independent presence, in others, a very weak presence, but in only a few instances is science policy well coordinated with other dimensions of policy. The problem is compounded by the fact that in many of the departments the economic analytical capabilities are extremely shallow. The CEA is not large enough to be able to do sectoral analyses, except in selected areas or on special topics. The departments must develop stronger capabilities, or other organizational arrangements must be made and capabilities developed. To begin to remedy this situation requires strong leadership from the Executive Offices as well as from Congress. And the latter, of course, requires that Congress understand the problem.

PART FOUR

GOVERNMENT-INDUSTRY COOPERATION

19 Innovation Strategies for Government and Industry
Umberto Colombo

The Problem of Innovation

The amount of industrial innovation in a country is one of its principal elements of vitality and also one of the most important conditions for the success of its economy. At the present stage of industrial development, innovation is of prime importance for Europe, since without it Europe's economic position would deteriorate as a consequence of the structural changes that are taking place in a rapidly evolving world economy.

Innovation involves a whole series of factors, not only of a technical nature, but also productive and commercial, that of economic, political, and social conditions. In certain cases the adoption of technical innovation inevitably imposes a parallel or prior occurrence of social innovation. Moreover, an excessive social stability may hamper interesting technological innovations.

The elements that not only open the door to innovation but indeed also stimulte it are technical achievements through research and development, entrepreneurial thrust, and of society's attitude toward innovation. European society, despite its wealth and the variety of its culture, and despite the turmoil largely initiated by the new generations, has remained substantially anchored to its past and fears the new, by which it has been so frequently seared in latter years. As a result of these conditions, the objective economic difficulties resulting from the scarcity of financial means and national resources have led Europe to adopt an attitude of defense and self-protection rather than more aggressive and risk-taking behavior.

A Comparison with the United States and Japan

For a deeper understanding of Europe's attitude toward innovation, we must analyze its causes. In this connection, it might be interesting to compare the behavior of other industrialized societies such as the United States and Japan.

Europe has a great tradition for scientific research at both experimental and theoretic levels and has been, and continues to be, quite inventive. However, Europe lacks that spirit of liveliness and the entrepreneurial thrust, downstream of invention, necessary to diffuse the innovations on the market. In contrast, in the United States, this spirit continues to be characteristic of the country, even if in recent times there are signs of diminishing innovation.

The United States is a young, enterprising, and lively country with a marked propensity to geographic and social mobility and with fewer rooted traditions than Europe. The United States looks to the future with hope and the idea of growing material wealth for all, which, until recently, has been the foundation of life. Although the traditional veiws regarding wealth, economic growth, and strictly material demand are subject to an evolution toward more qualitative objectives, there remains a fundamental optimism, a confidence in the future that is a spur to innovation.

The legislative and educational systems in the United States are structured to favor and enhance the value of entrepreneurial drive. Laws favoring market economy and competition include the antitrust law that protects smaller firms, which are thus in a position to compete with the more powerful companies, the disinvestment directives to the large corporations in order to favor free competition and prevent cartels and excessive concentrations, which appear to be natural in Europe, and the antimonopoly action to which companies like Standard Oil, duPont, Western Electric, and IBM have been and still are subjected so that they would not become absolutely monopolistic in the oil, chemistry, telephone, and computer fields.

Its position as a super-power imposes on the United States a major commitment toward advanced research in strategic sectors such as aerospace, computers, automation, and the development of conventional materials. Progress in these sectors induces additional innovation in other sectors of the economy. But the most positive element that has up to now favored innovation in the United States is the diffused belief in the superiority of a market economy and the adaptation of the various factors, including the educational and

Innovation Strategies for Government and Industry 211

the banking systems, to this belief. As a result, innovation in the United States is strongly market-pulled.

Unlike the United States, Japan faces constraints similar to those of Europe and, in fact, has serious shortages of raw materials and other resources, including space, that have created massive problems of an ecological and social nature such as tension and changing societal values. Japan, however, is not framed within a commnuity like the European countries that permits each country to optimize its own particular situation in relation to that of its neighbors. Incidentally, I should mention that the very effiicient functioning of the Federal Republic of Germany, and its hegemonic economic position in Europe, is at least in part the result of the integration, because of the Common Market, of its economy with the weaker economies of its community partners.

The geographically isolated position of Japan has caused it to pay a great deal of attention to its relationships with countries that supply raw materials and other primary goods and to which Japan must in turn supply its products and technologies. Japan operates in a market economy, but unlike the United States it uses its domestic market as a solid base for its productive expansion. Thus Japan practices a clear strategy toward foreign markets, toward which its industrial trading system behaves in a coordinated manner, unlike the United States for which market behavior abroad follows substantially the same rules as are applied at home.

In Japan, there is a close relation between the political and the economic systems, with an interweaving of relations between government, banks, and other financial institutions and industrial and trade structures. The economic and trade policies followed by the Japanese government are chosen in close contact and coordnation with industry. Unlike the American and European approach, this is not considered a fault but a positive connotation. Similarly, the legislation of the country is devised to protect and to favor domestic industry.

Since World War II, Japan has made precise strategic choices. Military defeat induced Japan to emulate the victor, in a spirit of humbleness, once its ambition of domination was overcome, and in the early post-war years, Japan copied as faithfully as possible external technologies and productive structures, particularly those of America. Japan then grafted its own innovations, which have taken on an increasing weight and depth, resulting from independent research, until it achieved positions of leadership in certain, even highly important sectors such as iron and steel, shipbuilding, electronics, instrumentation, watch-making, photographic, and motor cycle industries. Thus Japan has gradually lost its inferiority complex, overcome its imitative strategy, and, through its own inventiveness, become capable of turning situations of disadvantage into big market opportunities, largely as a result of industrial innovation.

The innovative capacity of Japan is particulaly obvious at the organizational and job efficiency level. A systemic approach allows coordinated action, a kind of team game, by various forces—political, financial, productive, commercial, and social—each of which respects the role of the others. This approach acts as a multiplier bonded to the structure of society. This unusual innovative performance is related to the specific culture of Japan and, possibly, also to the use of an ideological language and a corresponding ideographic script. The language for ideas, in fact, favors a software, rather than a hardware, approach, and a probing in the direction of the function for which something is thought of or produced, rather than in the direction of the specific technology possessed for its attainment, as often occurs in Europe, or of the market that is to use it, as frequently is the case in the United States.

The Japanese tendency toward a highly precise vision of things and toward miniaturized work initially favored penetration in such technologies as precision mechanics and such industrial sectors as watch making. The Japanese way of thinking, however, which is less bound to the type of technology and more directed toward the function of use, has led Japanese industry to pass on to technologies more suited to fulfilling the function of time measurement. The Japanese way of thinking, however, which is less bound to the type of technology and more directed toward the function of use, has led Japanese industry to pass on to technologies more suited to fulfilling the function of time measurement. The Japanese watch-making industry is gradually disengaging from traditional technology and adopting nonconventional solutions such as quartz oscillators and optolectronics, without the psychological attitude of a follower vis-à-vis the leading Swiss industry.

The extent of Japanese innovation is also correlated to the very high average level of education. The strong innovative charge characterizing Japan today also corresponds to the scale of values of Japanese society, which is simultaneously highly traditional and incredibly modern. These values are allied to a marked "esprit de corps" that goes from the enterprise level up to a deep-felt spirit of national solidarity. For Japan, this attitude is a necessity of life.

Japan also has a decided desire for revenge that, however, has been transposed from the military sphere to the economic.

The situations of the United States and of Japan are certainly not immune from danger. In the United States, in fact, where government intervention is often viewed with diffidence, the role of government in organizing social demand has been relatively minor, and industrial activity has prospered as a result of the remarkable mobility of the labor force and a comparatively high level of unemployment. All these elements have been decisive for the development of American economy, but the values lying at their base will soon need a deep revision. In Japan, instead, the still low but growing social and generational tensions may well loosen the tight knots of solidarity and accentuate the workers' claims. In addition, Japan has operated as the center of a vast geoeconomic system that not only provides goods and services in exchange for resources and raw materials, but that also exports manufactured articles that require a considerable availability of labor, whose cost is no longer so low. This situation, too, is destined to further deterioration in the future.

If we now examine the context and approach of Europe to innovation, we must acknowledge that other factors have acted negatively in addition to those previously mentioned. The great and diversified cultural wealth of Europe has not succeeded so far in becoming a valid instrument for innovation. Europe, moreover, has been conditioned by its history of glory and power, by its political and cultural heritage, by its strong traditions, which are hard to leave behind. Most of the European countries were victors in World War II—and it was Germany and Japan who reacted strongly and successfully in the economic field.

The postive approach toward market economy and the new, typical of the United States, is followed far less in Europe, which has left behind an experience of colonialism and of privileged markets. Past experience has brought forward a pronounced hierarchy of economic, social, and cultural values. European tradition leads to a greater social rigidity than the United States, and the social-climbing phenomenon is still much less common in Europe, where family tradition, name, and membership in certain groups continues to have an indifferent importance.

In Europe, laws and public attitude have not favored a lively and diffused competition, but rather have permitted the formation of monopolistic or oligopolistic situations in several industrial sectors. Similarly, industrial cartels of a defensive character, and therefore substantially noninnovative, have formed, contrary to what takes place in Japanese industrial and trading concentrations, which instead are directed toward attack abroad.

The European market proves to be considerably less stimulating for the diffusion of innovation than in the United States or Japan, but the strong scientific and technological tradition causes the technology-pushed type of innovation to be relatively more important. This appears obvious when considering the sectors in which Europe is strong, such as precision mechanics, optics, machine tools, pharmaceuticals, dyestuffs, and other fine chemicals. Other examples of technology-pushed innovation in Europe are the float-glass and the L-D oxygen steel-making processes, both of which have spread throughout the world; air transport by helicopters, which later found a greater independent development in the United States; sea transport by hovercraft, which has not yet attained a full market success; and, at the product level, polypropylene, from which the company that invented it has not obtained the best advantgaes in the marketplace, or carbon fibers, an unquestionable technological achievement, even if the innovation is diffusing in the market at a very slow rate.

In some sectors, this technological ability maintains the positions of strength attained through innovation—as in the case of pharmaceuticals, dyestuffs, and other fine chemicals—but Europe is not always capable of fully exploiting the potential of market development and sometimes is not even in a condition to defend positions of initial strength against continued innovative competition from abroad.

This situation is aggravated because the lack of political unity inhibits the concentration of effort and the strengthening of technological competence necessary to cope with dangerous situations in an innovative manner. This is the case, for example, of the Swiss watch-making industry, which had certainly foreseen the development of large-scale production and the consequent mass approach to the market, and later the introduction of electronics. However, the Swiss watch-making industry was not in a position to use these developments promptly because of an insufficient blend of internal skills. The Swiss watch makers are now compensating for this disadvantage, by lining up with the new technologies and integrating their previously missing skills. The extent of their success will depend on the adoption of the new technologies coupled with advertising and emphasizing the value of the Swiss name, traditionally still a guarantee of competence and capability. Perhaps, however, this major effort of technological and managerial adaptation would lead

Innovation Strategies for Government and Industry 215

to more rapid and effective results if Swiss industry were better integrated within a European framework.

Europe often follows innovation by others even when, in some cases, the invention originated in Europe. It is more difficult to be competitive in this situation. This is true for some branches of electroncs and informatics, as well as for precision mechanics, scientific and control equipment, and sensors.

Why is Europe often more inclined to operate in sectors where the continued thrust of innovation comes from other quarters, adopting a defensive attitude in relation to the aggressive strain of the countries with which it has to compete? To answer this question, it may be useful to refer to sectors, such as ship-building (undeniably European in origin), the heavy steel industry, and the automotive industry. In these sectors Europe has had to yield ground to Japan to an increasing extent, not so much because of strictly technological deficiencies but rather because of the great organizational ability of the Japanese and their general vision of production and commercial creativities. In addition, the Japanese worker, in common with his Amreican counterpart, is far more willing than the European worker to accept burdensome and hard-working conditions. Japanese patriarchal society guarantees the worker to the utmost, whereas the more individualistic American society offers the worker opportunities to find a new job and incentives to stimulate his commitment. Europe, instead, has a long tradition of social protection of the worker—a tradition that represents an essential value for its society —and the worker tends to become increasingly involved in the decision-making process through phenomena such as industrial democracy and "mitbestimung." This different situation may lead, at least temporarily, to lower competitiveness in European industry.

All this still does not explain the unsatisfactory innovative capacity of Europe and its tendency to follow the innovative process led by others.

The technological approach to innovation, typical of Europe, has resulted in the lack of other types of innovation, for example, innovation in marketing. For example, the Xerox process, although dependent on a technological invention, was accompanied by an equally significant marketing innovation, leasing the copying machine. This latter innovation has led to a new mode of use, switching the cost to the consumers from an initial investment to a day-by-day cost relationship. This type of innovation, which is based on an analysis of latent demand and which caused mass communications to explode in the business area, is typical of the United States, which is remarkably prompt at anticipating market needs and integrating technological innovation with marketing innovation, thus allowing the former to develop.

This readiness to innovate in relation to market needs, however, has aspects that should be considered with a certain discernment. It could favor a propensity toward consumerism and planned obsolescence, elements that might instigate crises since they are becoming increasingly less welcome to a more attentive and informed society. Furthermore, the European market cannot realistically be regarded as a true domestic sales outlet for European industries. The common market is certainly not, as far as the public demand is concerned, and Europe the "buy European" issue is practically nonexistent. In contrast, the American and Japanese markets are well protected and isolated. They actually represent big domestic markets readily accessible to the respective industries. Moreover, as far as public purchases are concerned, there is more coordination and certainly not the type of bitter conflict that exists among European governments that tend to lean toward their respective national industries.

A Strategy for Innovation in Europe

To modify the European situation in a positive sense requires a clear vision of intended objectives and the means necessary to achieve them. This implies the adoption of a strategy that Europe has not thus far been capable of providing. The problem is quite different for the United States, one of the two world leading powers with the structure and strategy typical of leaders. The strategy of other countries must be modeled on that of the leader and, according to economic circumstances and finalities, can be a strategy of defense or of attack, but always a strategy of coexistence.

In the immediate post-war years, Europe was engaged with its reconstruction and the benefit of American aid that favored the adoption of the American model in Europe. This kind of dependence is significant in relation to the pattern of adopted technology, which was quite different from the pattern Europe had been independently developing up to that time. The acceptance of a model of production based on large-scale units and parcelized work has left part of the European labor force unqualified.

Since the war, Europe has engaged in a series of endeavors, such as the development of the supercomputer, the supersonic civil aircraft, and the fast breeder nuclear reactor, thus pursuing age-long dreams of glory and power. Big projects were consequently conceived and organized. The lack of political unity and the divergence of economic interest among European countries, however, have been major factors leading to failure in the first two instances. The fact that Europe is not a political superpower may be detrimental to the third.

These examples indicate that it is not sufficient for Europe to merely have its technological papers in order. To achieve a stable competitive position, Europe must give itself a strategy and be ready to adjust it to the rapid evolution of the world economic and sociopolitical environment.

As we have seen, Europe is in a considerably more complex situation than Japan. In the first place, Europe is not politically united, and the economic integration process commenced in the 1950s has so far not been brought ahead to a sufficient extent. In other words, Europe is still facing the problem of how each of its member countries may fit itself into a framework of progressive European economic intergration, and this precludes the more important problem of how to optimize the situation of Europe as a whole in a world economic context. This means that Europe has been incapable of giving itself a unitary strategy, and that the process of strategic planning is more similar to that of a loose conglomerate than to that of an integrated and well-coordinated industrial enterprise.

We certainly cannot wait for the event of a European political union to arive at a European economic strategy, and neither, for that matter, would the setting up of a unity at a political level be in itself a sufficient premise for a valid strategy. What needs to be done however, is that tracing the outline for even a rough and limited converging strategy of the different European countries must account for the objective of a long-term political unity. In order words, a strategy of economic and social development in the different European countries must not hamper the objective of political integration.

Since Europe has a dramatic scarcity of natural resources, it is clear that its economic strategy and its very social welfare depend to a notable extent on the competitiveness of the industrial system. This in turn is conditioned by innovative capabilities. It is obviously unfeasible for each European country to innovate on its own account in many sectors. In fact, there already exists a more or less marked specialization of the different countries in different technologies and industrial sectors. At present, these specializations cannot be radicalized to arrive at a highly accentuated regional division of productive activities. This would be easier to achieve if Europe had united political and monetary interests.

Looking at the situation of Europe in relation to its interdependence with the other countries of the world, it is unthinkable that the process of an international division of labor could be allowed to proceed to the point where Europe becomes totally dependent on outside supplies. At least for the coming decades, a fundamental element of a European strategy must be the rationalization of the base industries, characterized by a relatively low unit value of the products and by processes that tend to be increasingly capital-intensive and concentrated in a few large plants. Rationalization in these base industries should lead to a better balance between centralized and decentralized production, the latter capable of compensating the apparent diseconomy of scale with a greater flexibility in production and hence a greater adaptability to market demands. Another element of rationalization is the distribution among the different European countries of these base industries generally classified as mature sectors. A third and perhaps more important element is the degree of dependence on imports that Europe is prepared to and can accept. This problem is particularly relevant to the issue of industralization of the less-developed countries.

Besides the base industries, there exist a number of industrial sectors for which one or another of the European countries is in a leading position at the world level. Conversely, each country is weak in other sectors, in the past having followed policies of safeguarding its own position even with no pretense of converting such sectors of weakness into sectors of strength. Viewed from a European standpoint, this policy toward weak industries certainly constitutes a strategic error, since it tends to disperse intellectual, entrepreneurial, and financial resources over a too broad arc of activities. A greater market specialization, at a national level, can and must be achieved in Europe, and perhaps it is easier to start in the less-critical sectors. This specialization should be favored by technological transfusions from one sector to the other, and from one country to the other, to increase the competitiveness of the different European countries in the respectively chosen sectors at a world level. Such a strategy should also be applied to those advanced sectors that do not call for an exceptional commitment of human and financial resources. When such a commitment is necessary, Europe will have to move from the outset in an integrated way, with a highly careful choice of the sectors in which to commit itself, avoiding sectors motivated essentially by considerations of prestige and not backed by solid economic prospects. Only a united Europe, with its major political weight, could sustain.

Going back to the more general problem of Europe's role in technologiaclly advanced sectors, the question is not so much that of Europe's presence or absence in a given sector but, rather, the nature of its presence and the related attitude. Europe, for instance, is certainly capable of developing its own strategy in the sectors of informatics, telecommunications, and space technologies without necessarily following the models set up by others, and therefore without yielding to the temptation of committing itself to a follow-the-leader type of research that most likely would produce negative results.

Clearly in the overall context of industrial policy, even an economically integrated Europe would not be in a position to maintain a valid competitiveness in too many sectors. When strategic considerations related to the safe and continued availability of indispensable goods do not prevail, Europe should decide upon and willingly accept a substantial dependence on other countries. This strategy becomes a necessity if Europe desires to recuperate the positive elements of market economy, namely efficiency, entrepreneurial spirit, and innovative thrust.

It must be acknowledged that in Europe a basically wrong policy of protecting weak sectors or the prepotence of oligopolistic forces, which have brought about the formation of producer cartels, have dulled entrepreneurial spirit and led to a de facto limiting of innovation. This situation must be firmly rectified by enlivening competition within Europe and possibly by joining the interests of European producers regarding export policies, following the example of Japan.

In cases where it might be necessary to temporarily retain cartels or other forms of concentration among producers in a sector, a role of control and supervision by governments and preferably by the community itself would appear indispensable.

Europe has a long tradition of sophisticated production calling for highly complex individual skills. This tradition, at least in some of the European countries, has flagged recently for a series of reasons in parallel with the process of concentrating production in larger plants. This trend should be inverted and could be achieved through a new decentralizing thrust, connected, on the one hand, to better regional management including the safeguarding of the European cultural heritage and, on the other, to the necessity of reducing unemployment.

The requirement, previously refered to, of reinforcing in Europe the positive elements typical of market economy must not give rise to a push toward a laissez faire type of economy that would lead to increasing disparities and internal contradictions. On the other hand, a rigidly planned economy does not encourage an innovative attitude that can only be manifested through a greater participation and involvement at all levels, and results, instead, in a bureaucratic top-down, rather than a bottom-up approach, thus hampering entrepreneurial spirit. Therefore, it is necessary to attain a harmonic balance between characters of a free market and those of a guided economy.

The European Role in Socially Oriented Sectors

The European tradition of labor protection, its policies of low unemployment, the welfare state and related social infrastructures, its performance in urban planning, its vanguard position in promoting a different and less alienating way of producing in industry, the close relationship between government and industry, and the extent of direct political intervention in the national economy to achieve a better adaptation of industry to social requirements should place Europe in a favorable position vis-à-vis socially oriented innovation, the need of which is badly felt in modern societies.

The increasing interaction between agriculture, industry, and services, which is common to all advanced societies, is taking place to a considerable extent in a Europe pressed by such problems as the scarcity of natural resources and high population density. This proecss of gradual convergence of the productive and service structures offers industry the opportunity to play a primary role in fulfilling important objectives of society. The innovative capacity of industry and its organization—flexible enough to face different and complex situations—make it a suitable instrument to cope with the problems posed by a highly interconnected system of activities. That the direct goal is the satisfaction of social demands and that related expenditures are substantially part of the public budget make it necessary for governments and the public administration to control and orient the action of industry in these sectors. Industry must understand this function of public power and cooperate without pretending to be capable of doing everything by itself.

These socially oriented sectors cover a very wide range of activities and involve problems that are by no means trivial. It is once again necessary to reflect on the potential benefits that Europe could obtain by adopting a coordinated approach, that is by gradually replacing the public demands of the individual countries with a much stronger European public demand. This would allow a broadening of the market base for the related goods and services and would favor a higher degree of specialization and professional content, hence of innovation, in the industries concerned.

It is clear that to give the state a more important role as a purchaser of industrial goods and services means to reduce the volume of expenditures that the citizen can afford on a private basis. This should entail the outcome of a more austere way of life. The reduction of the level of private expenditures, however, could prove to be less dramatic and smaller than might be thought. In fact, the deprivation of potentially available private goods is at least in part balanced by the reduction of social costs deriving from the new mix of economic activities. Moreover, this deprivation may be more easily accepted if it concerns products that have given rise to an excessive consumerism with related waste of resources.

Furthermore, one should not forget that these socially oriented technologies and products are indeed exportable either as such or after adaptation to different situations and needs. There is, in fact, a growing demand, particularly in the developing countries, for goods and services related to social infrastructures (schools, hospitals, public transport, housing), and this could provide Europe with additional income and contribute to socioeconomic world development. It is important for Europe to maintain a dynamic pace, however, if the potential advantage of its higher propensity toward socially oriented innovation is not to be lost in favor of more aggressive and market-oriented competitors.

An analysis of socially oriented industrial activities leads to the consideration of the direction followed so far by industrial innovation in general toward the expulsion of the labor process from the productive process. In fact, the concept of an ever-increasing productivity of labor, favored by technological progress and innovation, has been traditionally associated with the idea of an expanding and growing economy. In conditions of scarcity of financial means and natural resources, the assumption of an ever-growing economy, where the resource represented by labor is relatively scarce, is seriously challenged. If even during periods of sustained economic growth the level of unemployment has increased in industrialized countries as a result of continuing productivity increments, unemployment could become more serious, at least during the next decade, in view of a slower rate of economic growth. A crucial issue in this respect is the understanding of the relationship between innovation, from the initial discovery to wide diffusion in the market, and unemployment.

As indicated by Christopher Freeman, in the initial stages of innovation associated with key patents, many prototypes, intensive research and development for new products and applications, little standarization and small-scale production, jobs are created as a result of the innovative process.[1] On the contrary, in the later stages of innovation when the goals of research and development have shifted largely to cost-saving, most applications have already been developed, products are largely standardized and manufactured on a big scale, and labor is displaced.

Although it is historically true that over the lifetime of a basic innovation more jobs may be created than displaced, the problem has different connotations when considered in light of the international economic context. It is clear that those countries that do not contribute sufficiently in the earlier stages of basic innovations are forced to import unemployment in the hidden form of the final stages of innovation. Under this aspect, innovation, called upon to solve certain socioeconomic problems, may at the end create new ones in the form of unemployment. This is clearly an oversimplified discussion of the impact of innovation on employment. For example, the relationship between capital and labor and the degree of their substitutability or complimentarity should be further investigated and included in the picture.

In the future, technology should become appropriate to the changes in factor proportions, at least in the medium-term surplus of labor and scarcity of capital. This change in factor proportions could serve in itself as a stimulus for new and different innovations instead of artificially, if not forcefully, increasing labor intensity in existing activities where a surplus of labor leads to uneconomic results. This leads to the problem of identifying and developing technologies appropriate to the European economic and social system, where the abundance of labor force is not compensated for by a low unit cost of labor. Several industrial sectors lend themselves to the development of appropriate technologies economically viable in a European context:

- Using local mineral resources to manufacture building materials.
- Using scrap material.

- Using waste for different purposes, including the production of energy.
- Using various sources, including the minor ones for decentralized energy systems.
- Using different forms of energy produced (total energy systems).
- Using biological concepts and methods within chemical and agricultural production.

The development of appropriate technologies in all these sectors could lead to productive activities characterized by higher labor- and lower capital-intensities.

These appropriate technologies must not be seen as counterposed to the traditional ones, but rather as complementary to them in the context of a better balance between centralized, capital-intensive and decentralized, labor-intensive activities. The real objective is to pursue an appropriate combination of technologies within a strategy based on what may be called technological pluralism.

Europe's great shortage of resources may be conveniently faced by considering, in addition to the use of imported conventional raw materials, the employment of alternative local resources, of substitute and more plentiful materials, of a developed recycle economy, of a different way of designing, and of producing goods to lengthen their life and allow ready replacements and repairs. Appropriate technologies and technological pluralism thus represent a complex and articulate pattern of production that must be conceived and understood at a system level. In this sense, they provide a further notable opportunity for innovation and entrepreneurship. They also appear as an opportunity for exporting technologies more suitable, after due adaptation, to the real needs of the developing countries.

Recommendations

We have so far outlined the elements of a possible European strategy in which industrial innovation would play a major role. The creation of a climate favorable to industrial innovation is a real issue, an issue that has been analyzed several years ago within the Organization for Economic Cooperation and Development (OECD). The Pavitt report (1971) that presented the conclusions of such a study suggests the following objectives for a governmental policy in favor of innovation:

- Ensuring industrial competition, as the main pressure for technological innovation
- Ensuring equitable rewards for innovations through tax systems
- Ensuring that regulations, codes, and standards take account of both the social costs and benefits of the innovative process, as well as the flexibility and pluralism required for successful innovation
- Having active regional and manpower policies to deal with the changes in industrial and skill patterns brought about by technological change
- Using government procurement to upgrade the technical level of industry and to couple technology more effectively to collective social needs
- Encouraging the mobility of scientists and engineers, especially in and out of government laboratories
- Identifying policy measures to encourage science-based entrepreneurship
- Ensuring continued trade and capital liberalization, thereby heightening the pressures and incentives for technological innovation in all member countries, and maintaining the rapid, international spread of the benefits of new technology.

Most of these objectives would be valuable now in implementing a European strategy for innovation. However, it is fair to say that the present international economic situation is more complex and difficult than when the OECD report was written. In particular, the financial difficulties in which many industrialized countries find themselves have weakened the market economy and have led to a more direct and determinant role of governments. Therefore, more emphasis should now be placed on the creation of active policy instruments, capable not only of promoting actions in the direction of public and socially oriented activities, but, even more importantly, of adding vigor to the spontaneous entrepreneurial forces of the economic system. Such instruments should, on the one hand, act as incentives and adequately reward innovation and, on the other, discourage and punish enterprises that base their success on the creation of barriers limiting an active competition and unduly exploiting the market.

Their elimination of industrial cartels must take place in the context of a clear industrial policy at a European level. If, for strategic reasons, a sector has to be protected, this is not a good reason for avoiding intra-European competition. Rather, the protective measures must be adopted at the boundaries of Europe and be motivated by serious and defendable arguments. This would put an end to the present system of aiding the weak industrial sectors that each country carries on with its own criteria, and in some cases with a short-sighted approach.

A buy-European policy is needed on the part of the governments of the various countries, particularly in the technologically advanced and in socially oriented sectors. The government as a purchaser must not adopt a bureaucratic approach, but should instead aim at product quality and efficiency, thus stimulating innovation.

The medium-sized and small industries have traditionally played an essential role in the generation and diffusion of innovation. Adequate instruments to support the action of the innovative smaller enterprises must be devised. Today instruments aiming at this objective are operating at a national level in several European countries. So far no instrument of this kind is available at a genuine European level. A venture capital enterprise (EED) formed about 10 years ago on a European basis following the example of a successful practice in the United States, has failed largely because of the traditional attitude toward risk of the financial stockholders. A more recent initiative in this direction, SCIENTA, is perhaps in a better shape, although one cannot refer to it as an ongoing success. In any case, there seems to be a need for a European venture capital enterprise, managed with an entrepreneurial, risk-taking attitude and with adequate technological foresight. If the European banking system is not available for such an initiative, and if the large industrial concerns are not interested in being the stockholders of such an enterprise, the EEC could exert a promotional action and perhaps even sponsor the undertaking at its early stages.

The big industrial companies should also catalyze and sustain innovation within the smaller firms by creating laboratories or experimental centers for the transfer of technology, aiming at providing technical assistance and at stimulating the innovative effort of the assisted firms. Such laboratories could receive a public incentive, in view of their not strictly promotional scope. Similarly, governments should encourage the professionally worthy sectoral research and development associations, which are often the cooperative effort of many smaller firms.

As far as large innovative projects are concerned, often requiring major scientific and technological effort and a heavy financial commitment, a selective and attentive policy should be set up at the community level, illustrating the usefulness of installing a European Office of Technology Assessment, as suggested by Europe plus 30. Among actions that would certainly contribute toward stimulating innovation in Europe, those favoring transsectoral and international mobility of skilled individuals should not be overlooked. This issue of mobility is one of the main elements of a sound manpower planning policy.

Finally, industrial innovation is the product of the intellectual effort of capable individuals. In latter years, the attitude of people, and of the young in particular, toward science, technology, and industry has become more critical and suspicious. In fact, much of the unrest and difficulties of our time is ascribed, without a critical analysis, to science and technology, to the point of assuming in certain cases an a priori attitude of total rejection of technological innovation.

Industry in the past has made many decisions without a sufficient sense of social responsibility, considering that its main duty was to increase the wealth of society. In doing this, industry has sometimes exploited the lack of knowledge, within public opinion and administration itself, of possible unwanted side effects of industrial production or products. Now industry must realize that the assessment of technology should be incorporated as early as possible in its strategic decisions, and that a frank dialogue with the educational system and public opinion is in its own interest. This dialogue will provide a better understanding of industry's viewpoint and of the role that science and technology are now, more than ever, required to play in answer to today's problems. An action of this kind should also favor recruiting, for industrial research, young individuals, endowed with the necessary ability and motivation.

These and other general recommendations, in order to be implemented, need a thorough evaluation of their institutional feasibility and of the likely consequences of their implementation. It seems advisable, therefore, if the EEC is to play a concrete role in this issue, that a project on industrial innovation in Europe be launched and that close links be established with the directorates dealing with scientific and technological policy and with industrial and economic affairs.

20 Aligning Industry Planning and Government Policy

A. E. Pannenborg

This paper deals with the influence that government policy and measures can have on the program of research and development carried out in a private enterprise. As far as the latter is concerned, I limit myself to the pure private enterprise and do not refer to enterprises in which the government, through ownership or a major participation, might exert direct steerage. Government policy in Europe refers to a set of individual national policies, because until now the impact of the EEC in this subject area has been very modest. I come back to the EEC at the end of this paper.

It seems proper to approach the subject by first describing those chapters of government policy that affect industrial research and development. I do this in a phenomenological way without comments on the merits or disadvantages of specific concepts.

Government Policies

The oldest government measure with direct implications for industrial research and development is undoubtedly the patent law. In fact, patent laws came into existence before the advent of industrial research and development. Originally these laws provided protection to the individual inventor giving him the possibility of reaping the financial benefits from his own brainchild. These laws were applied in a directly similar way to the legal entity constituting the enterprise and have been since then an essential cornerstone in the world of research and development, as carried out within industry, though its practical importance differs from industry to industry.

Another classical field in which government policy has consequences for research-and-development activities in industry is that of technical education. In fact, the creation of the Ecole Polytechnique and of the first Technische Hochschule were pioneering concepts of which the consequences can be demonstrated even at the present day.

If we now focus on the question of how in modern times the government, as the means for political expression of the nation, has influenced the tasks and programs of research and development in industry, we can ascertain three broad areas:

- Collective needs of the nation (government as a customer)
- Safeguarding health and safety (repression)
- Industrial policy (mainly promotional)

The way in which the government policies in these areas influence industrial research and development is different for each area as is indicated:

Government Policies

	Affecting but not aimed at industrial innovation	Toward industrial innovation
Collective needs	+	+
Safeguarding health and safety	+	
Industrial policy		+

This diagram indicates that industrial research and development in certain cases has to react to government measures, whereas in other cases government measures are aimed at industrial research and development usually in a promotional manner.

In modern society, at least in Europe, the government has been empowered to provide certain services, often on a monopoly basis. Examples of such services are P.T.T., railways, electric power, water, and gas. Another specific kind of service is national defense. In these fields the government plays primarily the role of a customer, and because of the size of the needs, an important customer. The products required from industry to provide these services will in many cases be developed by industry on its own initiative; in other cases, governmental service will stimulate and sometimes fund specific developments within private industry. The interaction between industry and these services in the field of research and development is usually quite effective because the size of the organization permits a high degree of expertise within the service.

The next area, government measures for safeguarding the health and safety of the individual citizen, has quite a different character than the former. It entails a sizeable body of prohibitions and mandatory procedures, to which new products and processes of industry are subjected. It means that the design engineer in industry must realize the desired specification as an economic optimum not only subject to nature's, limitations but also subject to a list of man-made limitations. Recent years have produced an appreciable extension of these limitations. The scrutiny of new pharmaceuticals is much more severe than 15 years ago; protection against pollution, noise, and other environmental factors, has become much stronger; and legislation to protect the privacy of individuals against computer data banks will be soon forthcoming in many countries.

Finally there is the area of industrial policy within which governments often include measures to stimulate innovative work and research intensity in private industry. Such measures used to have a general character, working with fiscal means, but specific measures such as explicit funding under a development contract with the individual enterprise are sharply on the increase.

If we look back on the three areas just described, we see that governments have a wide choice of mechanisms at their disposal with which to influence or promote research and development in industry. In listing the most important ones, we come to the following means and methods: procurement, legislation, product approval, process approval, government or semigovernment research-and-development establishments, fiscal measures, and funding of research and development in private industry.

Effects of Government Policy

Standardization, which generally is of vital importance to industry, is often under the influence of national policy. The economic vested interests have lead in a number of cases to significant differences in standards between one country and another. It has then the effect of creating nontariff trade barriers and means a complication for the international enterprise insofar as the products to be offered in one country have to meet different specifications than products offered in another country.

In general, government measures taken on a national basis affecting or promoting research and development in industry, contribute to the distortion of free competition on the international scene. This certainly holds for explicit funding of development projects, but occurs—in the negative sense—also when repressive legislation in a field is introduced in one country at an earlier point or with more severe requirements than in other countries. If we look at the effect of direct funding, we find that it can be enormous. Notably, in the United States there are multibillion-dollar companies that receive up to 60% of their research-and-development expense under contract from the government. This forceful mechanism in the United States undoubtedly was one of the factors leading to the discussion of the so-called technological gap.

The requirements for product approval have risen very steeply in certain sectors of industry, such as the pharmaceutical industry. The expectation can hardly be avoided that as a consequence the rate of innovation in the affected branches of industry is bound to slow down considerably in the years to come. A completely novel situation was created when legislation introduced certain requirements before the technical feasibility of these requirements had been proven as in the instance of the legislation defining the composition of exhaust gases of automobiles in the United States.

In considering the various effects and consequences of government policy, one can only hope that these indeed reflect the present and future needs of society. Furthermore, it is evident that through this strong interference government is beocming increasingly responsible for the properties of the goods supplied to society.

The Policy of the Private Enterprise

Certain enterprises make a business out of doing contract research for third parties without or uncoupled to manufacturing operations. This type of enterprise is not discussed here. Instead, we concentrate on the enterprises within the manufacturing industry carrying out their own research and development for new or improved products or processes. Such enterprises will have an interest in maintaining their research-and-development capacity in a more or less fixed relation to their production capacity.

In general, the management of a private enterprise has the desire to minimize direct government influence on its operations and on its policy and strategy. In these years of increased direct government funding of research and development in industry, it is important to observe that economically speaking the timely creation of a new demand through government legislation or decision is more important to the company than financial support for its research and development. Another aspect of equal importance is the nature of the relation between a funding agency, and the ultimate customer of the product that will emerge from the funded research and development. This was seldom a problem in the case of research and development for defense purposes, because the funding agency and the purchasing agency, if not identical, belong to the same branch of the executive. Recent years have shown a rapid increase of governmental research-and-development funding for civilian purposes. Here the identity of the close coupling between funding and purchasing entities is not at all a priori guaranteed. It might be, as is the case in some countries in the field of nuclear energy, that research and development is carried out and funded by the central government, whereas the purchasing responsibility rests with a lower governmental level or with rather autonomous semigovernmental enterprises. In other cases governmental funding of industrial research and development is prompted by the aim of promoting advanced high-technology industries and might be applied to product research for which other private companies or even the private citizen is ultimately the customer. In these cases, the appraisal of the chances for ultimate commercial success should be carried out by the enterprise. The government, after all, is not equipped to carry this out.

Specific Situations Confronting Industry

The first category to discuss concerns legislation aimed at better protection of the citizen. If new legislation is under discussion and is to be set-up, industry should be consulted. The technical expertise involved is generally available at greater depth within industry than in government. The research-and-development organization of the enterprise is wise to anticipate the possibility of further restrictive legislation, not only in order to get acquainted with all aspects of the subject, but also in order to work out at a timely moment the changes in technology that might become mandatory under the new law.

Whereas in the former case the enterprise has no choice but to comply once legislation has been enacted, in the next category the freedom of choice for the enterprise is much larger. This is the situation when a new demand of society is opened up through the appropriate government decisions. It might involve additional equipment for motorcars or the start of a new broadcast service. Again the research-and-development organization of the enterprise should acquaint itself with the basic knowledge required sufficiently early and use this as input to top management of the enterprise who must decide whether to engage in the new field or not. Because of the layer structure of industry the foregoing holds at least as much for the components and materials industries as for the end-user product-industry.

We now turn to the possibility that government funds become available for sponsoring research-and-development work within industry. For those branches of industry where such finding is not an exception, it is important for management to define its aims and to derive from them a suitable pattern of behavior toward government-funding possibilities. Most governments, at least formally, are subject to budget approval by a legislature once a year, so that guaranteed continuity over several years can be a problem. A consideration often used by an enterprise confronted with the possibility of obtaining government funding is the danger that these funds would go to direct competitors if the company hesitates to decide. It is my experience that this consideration is often overemphasized with the inherent risk that the company is seduced into going outside its own field of true competence or commercial interest.

A feasibility study is generally attractive, and because of the ensuing influence on the concept of the products or systems it appreciably enhances the chances of the company to become a major supplier later. Finally, some governments allow risk-bearing credits to be given to industry, on a 50/50 basis for projects, the specification of which lies completely with the enterprise.

The Role of the EEC

Until now the member-states of the EEC have experienced great difficulties in formulating a community industrial policy. The industrial policies of the member-states themselves carry strong features of mutual competition and are, therefore difficult to reconcile. A particular point where the Treaty of Rome is not fulfilled and where accordingly the European Commission has repeatedly reacted, is the procurement policy of the governments of the member-states, which is generally fairly nationalistic.

The EEC has been compelled to refrain from an active industrial research and development-promoting policy in those areas where member-states have pertinent policies themselves. In those areas where no objections against funding from a European source are met, the community rules sometimes call for joint application by two or more enterprises located in different member-states. This rule, though understandable, is very objectionable from the point of view of the private enterprise, because it generally would imply the teaming up with a direct competitor, which would be a major policy decision not in proportion to the modest amounts available for funding from Brussels.

The EEC is contributing to European legislation, by standardizing requirements for certain products. This, undoubtedly, is welcomed at least by those industries that operate on more than one national market within the community.

Final Remark

For a believer in the system of free enterprise, the increasing impact of government on industry is not always regarded as a positive fact. One has to recognize, however, that the vastly increased complexity of society calls inevitably for regulatory power in certain areas, especially when the required economies of scale can no longer be achieved within the national boundaries. We from private industry should also realize how poorly government is equipped to deal with industrial questions that require a highly specialized expertise. The most realistic position, perhaps, is to accept the increasing interaction in order to be prepared to encounter the situation in the most fruitful way for the enterprise.

21 National Science Policy and the Private Sector

Herbert I. Fusfeld

My intent here is to provide some general comments on the nature of science policy and the emphasis it appears to be taking today in the United States. These perspectives are derived from the evolution of science policy at different periods in the United States. This evolution has brought about a necessary interaction between government science policy and the private sector.

Background

There is no clear understanding or agreement on what activities make up science policy. For simplicity, we should at least agree to use this term to represent the more accurate, but awkward, term "science and technology policy."

As a first consideration, science policy refers to an activity of government. It may affect the actions of nongovernmental institutions and personnel, but the term describes only the policies and actions of government. The separate decisions on expenditures and choice of technical effort that dictate the programs of university research faculties or of industrial research laboratories are conducted within the environment of whatever government science policy exists. When there is no explicitly stated government policy to influence or guide nongovernmental technical effort, and the objectives and pressures within the nongovernmental institutions are the prin-

---

This paper is based on work performed for the Organization for Economic Cooperation and Development, Paris. The views expressed are those of the author.

cipal or sole bases for allocating research-and-development activities, then that sitaution is the national science policy.

The obvious point to make is that there is always a science policy of government at any period. We never go from a period of no policy to one where we create a science policy effort. Rather, we are constantly changing our science policy to meet the changing needs of our society and economy.

To consider the effects of these changes more specifically, we can view science policy as composed of two broad characteristics, passive and active. In general, these correlate with concerns of science as an end in itself versus concerns with the national objectives. Expressed differently, national science policy can be considered in terms of:

1. Concern with the scientific and technical base of the country (i.e., the state of knowledge, research capabilities, and technical manpower)
2. Concern with the deliberate use of science and technology to achieve, or provide options for achieving, national objectives related to economic growth, foreign policy, national security, or quality of life

The implementation of a national science policy can be both direct and indirect. That is, the government can:

1. Direct resources of money and personnel to particular national goals, including that of adding to our reservoir of basic science
2. Establish the boundary conditions for the operations of both government and nongovernmental sectors that will affect the nature and quantity of technical activities and their applications

Both direct and indirect actions are present in periods of either passive or active science policy. This is easily seen by a brief review of past developments in the United States.

Past Developments in the United States

Science policy in the United States was largely passive, with a few notable exceptions, until World War II. The period of World War I was one of those exceptions, and we can also identify ongoing concern with agriculture and public health as specific points of emphasis for active support in periods of an otherwise generally passive science policy. During World War II and up to the present, there has been a sharp change to activism in science policy.

The principal determining factors for government policy and action relating to science and technology are:

- The perception by government (including interactions between an administration and the general public) as to how science and technology can play a role in achieving some national objective
- The determination by government that the actions of the nongovernmental sectors are not adequate to supply that role for science and technology without some actions by government

We can draw upon these statements to provide a grossly oversimplified summary of United States science policy during our 200-year national existence. The broad evolutionary phases, all of which exist today, are:

1. A steady growth of government support for the underpinnings of science and technology such as education, basic research, and, within government laboratories, measurements and standards. The nongovernmental sectors were relied upon to provide for the use of science and technology in the economy and any other perceived national needs. This was essentially a passive phase concerned with strengthening science and technology generally rather than with the pursuit of a specific endeavor.

2. Specific areas were identified as requiring government actions to ensure and expedite the optimum role for science. These included agriculture, health, and aviation, all largely pursued within government laboratories. The pursuit of these mission-oriented areas formed the nucleus for an active science policy.

3. The requirements of World War I set the pattern for a major form of activism in science policy that bloomed with World War II and continues today. This is the active government role in pursuing the development and applications of science and technology intended to support major national objectives in which the government is the principal customer. National security is clearly first, but atomic energy, space, and others are included.

4. Finally, there is an increasing government involvement in science and technology efforts intended to support major national objectives in which the government is *not* the principal customer. Energy is an obvious example. Others are materials,

transportation, and housing. And major attention is going to the role of technology and innovation in economic growth. Pursuit of all these areas defines a truly comprehensive active science policy by government today.

The passive science policy of earlier days was easily understood and accepted in regard to the role of government and the nature of its actions. This is generally true of the passive aspects of today's science policy, though controversy exists over the level of effort and the effectiveness of the mechanisms used to distribute funds. These comments refer primarily to support of basic research at universities and certain technical activities within several of the government laboratories.

As science policy shifts toward an active phase, encompassing the civilian sector, many more questions arise. The two most serious and fundamental ones are: How does one decide which particular technical efforts to support through government action? and What mechanisms or transfer processes should be used to provide for effective use of the science and technology supported by government action?

The answers to both questions must involve the private sector. Thus we have the paradox that the more active the government's science policy, the more it must depend on the private sector to achieve optimum effectiveness. Stated differently, an active science policy pursued without adequate involvement of the private sector acting in accord with its traditional procedures will not be effective.

These points are simply stated here. The basis for the statements can be seen in a brief consideration of the industrial research effort.

Role of Industrial Research

I emphasize only those several aspects of industrial research that must be considered in the development and implementation of a national science policy. This rests upon an understanding of the role of the private sector itself.

In order to focus quickly on the few points I make, consider that the private sector in Western market economies has the primary responsibility for the manufacture and distribution of goods and services. The operation of the private sector in performing these functions possesses certain characteristics relevant to our discussion:

- It decentralizes decision making. Without minimizing possible trends toward concentration, the fact is that each individual firm makes decisions concerning the use of only those resources within its own control, and there are a great many firms.
- The private sector allocates the use of these resources to produce maximum return on investment consistent with long-term growth and stability. This return, or profit, viewed over the long term is a measure of the efficiency with which resources are used, given the prices set by society on labor, capital, energy, and so on.
- The optimum use of resources provides an internal pressure within each firm to maintain a proper balance among those resources (i.e., the level of manufacturing facilities, distribution, research and development, and all other components must be compatible with each other and with the financing and management strengths available to the firm).

Now consider the role and characteristics of industrial research in this environment. It is not an independent activity, set apart, feeding occasional ideas and technical breakthroughs along a one-way communication link to an eager and waiting production line. It is, and must be, very much a part of the industrial system just described. In accordance with this, plans should be made for technical programs relevant to the business strategy of the firm, a level of technical effort compatible with the needs and abilities of the firm, and mechanisms for conversion to use of successful research-and-development programs.

The key to all these activities, unique to industrial research, is balance. This is both qualitative as regards business strategy and management capabilities and quantitative as regards use of available funds and manpower with consideration of the probable return. The process by which one arrives at these judgments includes the cost and technical feasibility of adapting successful research and development to workable manufacturing processes using economically available materials to give satisfactory performance in use.

There are, in short, certain disciplines that make up industrial research. This is why a project concerned with removing sulphur from coal, even with the same outline of technical tasks, has fundamentally different attributes when conducted in industry as contrasted to a university or government environment. Among these attributes are:

- Consideration of the system used to develop manufacturing processes and arrange distribution
- Consideration of all technical characteristics of a final product or process as defined by the needs and constraints of the user
- Consideration of the interactions among market demands, cost, investment, and technical performance of the product or process
- Consideration of all options available to meet broad needs of the potential user with regard to a specific mission-oriented objective, either through competitive technical approaches, substitutes, or nontechnical approaches such as the use of economic incentives or penalties

The industrial research community, particularly the research manager, is the bridge between science and the user that must account for the transfer process between the two. The research manager is aware that the function of research and development is to provide options—for solutions to problems or for investments—that are acceptable economically to society. The research-and-development activities are integrated with a complete manufacturing and distribution system, and all parts of the system are involved in the earliest planning and the ultimate use.

These are, of course, precisely the approaches required to answer the questions raised regarding an active science policy in the civilian sector: how to decide on priorities among technical programs and how to provide for effective transfer. The mechanisms for both reside within industrial research. The problem is how to couple this know-how with government programs.

Government–Industry Interactions in Science Policy

Let us now return to the needs and implications of an activist science policy in the civilian sector. Each mission-oriented area must be identified separately. There is little in common among the import problems of shoes and textiles, the national and commercial requirements of ocean mining, and the design of nonpolluting automobiles. No common policy or approach applies to all. Rather, the sum of government programs in this field will make up the civilian-oriented science policy.

By treating each item separately, we can now ask a more fundamental question than how to choose priorities and how to use the results. This is: Why? Why should the government become involved at all in the technology of a particular industrial sector? What, exactly, is the problem in that sector?

Here it is absolutely vital that the industrial research community be asked to work with appropriate government agencies in advance of any major government involvement. One must examine carefully the balance within an industry and within individual firms. If restraints of markets or finance make it desirable to reduce research and development in a particular area, a forced increase in research and development will not result in greater exploitation. There are, obviously, areas in which national needs may dictate a more accelerated use of technology than the private sector can justify on purely economic grounds. Again, the industrial research community can contribute to the identification of how much acceleration can be accomplished through direct support of technical efforts and how much through nontechnical actions.

There will be areas where it is not economical for the private sector to conduct research and development either because the market may not justify the probable costs, or because the required technical effort is too large for private resources. Both situations define an inherent imbalance between research and development and the ability to convert it to use . The function of the industrial research community in these cases is to provide recommendations to minimize the mismatches by involving the private market development capabilities in the first case, and so structuring the technical program in the second case that effective involvement of private sector research and development can aid the transfer process.

In other words, there are necessary inputs from the industrial research community in the planning, conduct, and exploitation of any active government programs intended for the civilian sector. The intent here is not to recommend mechanisms but to bring about serious efforts by the government to obtain these inputs. Further, there must be a recognition that the value of involving the industrial research community can be both positive, to ensure effective planning and use, and negative to avoid government action where none is called for.

It is also critical that the industrial research community be involved not merely as technical advisers who have an industrial background, but as active members of a manufacturing and marketing system. That is, there should be official recognition of the role played by the private sector and an acceptance by government of the need *and* the desirability of working through the private sector in the development and implementation of science policy.

Current Pressures

Much is said today about the "new economic context." Certainly the subject matters of interest to government in different eras change. Today, several large concerns of a technical nature are energy, materials, and environment. The three are closely interrelated.

What may be considered new in nature, as well as in subject matter, is the increasing desire by government to draw upon science and technology as tools in the general objectives of stimulating the domestic economy and strengthening its position in foreign affairs. There have always been specific examples of such association, but there does seem to be a more general recognition throughout government, at least intellectually, that there are contributions to be derived from a science policy directed toward the economy and foreign affairs.

This, then, is new. The several principal subjects that should receive considerable emphasis in an active science policy are themselves deeply intertwined with both domestic and foreign policies. The results of programs in these three areas have a major impact on production costs, on the availability of capital for economic growth, and on our self-sufficiency in the world.

So we return to the main thrust of these general comments on science policy. The increasing concern of society to use science and technology in civilian-oriented pursuits, domestic and international, has focused government science policy on these objectives. But such efforts must be transferred through the private sector, and must involve this sector, particularly industrial research, throughout the development of any active science policy so directed. The challenge to all of us is not, therefore, how much government money to spend on what, but how to bring about government–industry cooperation as a matter of policy, in contrast to the adversary roles often taken today. This should be embodied preferably in legislation, hopefully in the public attitudes of congressional committees, and certainly in the conduct of those in the executive branch responsible for establishing and carrying out science policy.

22 Technology and Psychology: A Query

Gerhard Mensch

Is the recent technological stagnation in the industrial societies a psychological problem or one of economic or technological origin? Unfortunately, this question has not received the level of attention that it should.

The problem is highlighted by recent figures issued by the Organization for Economic Cooperation and Development (OECD) concerning the so-called investment gap in various industrial countries. Some industrial countries invest far less than others so that a wide variance exists. If the less-investing industrial countries are to catch up with the average in the next few years and if this tremendous gap is to be overcome largely through investment in innovation, then an enormous amount of human energy is required.

The need for organizational and technological change would be so extensive and the innovative processes would need to be so strongly intensified that obviously the creation of new technology is not the bottleneck, neither is financing. Rather it is motivation, anxiety, courage, and nerve. Therefore, I hesitate to trust too much in science and technological policy when dealing with the socioeconomic problems of technological change in the next few years.

On the other hand, economic policy makers seem not able, or ready, to deal with structural policy issues. For them, the problem is obviously not very handy since it requires many projects of "makeable size" to be performed by "makers." Because these ought to be innovative projects, economic policy makers have no means of dealing with the multiplicity and heterogeneity of these problems.

Therefore, since macro-policies are likely to be ineffective when creating a sufficient rate of major innovations in new branches of the economy and improvement innovations in existing lines of business, we must expect market mechanisms to do the job only at their own pace. That means that as long as the problem seems to be that enormous, private individuals and firms will not move until many others seem to move, that is, until many "makers" see new technology feasible and "makeable" in their scale of operation, small enough in scale and low enough in risk to be operational. Obviously, that involves so much cooperation between market and nonmarket institutions that the rate of innovation in the next years will be lower than desirable if there is not a change in the psychological climate.

History teaches us that in times of economic depression the economic system becomes eventually ready for a spurt of major innovations. Until that time, however, it is easier for firms to find appropriate technology for their internal innovation purposes, namely rationalization and replacement of existing production processes and distribution systems. External innovation targets, namely those oriented at new goods and services catering to shifted needs of consumer and clients, expansionary new investment opportunities, will be less often found and taken. Therefore, I assume for the next few years that trends might continue as they have in the last few years with respect to imbalanced technical change. As we found out by classifying these innovations collected in the science indicator report, between 1960 and 1973 the ratio of expansionary versus modifying innovations dropped alarmingly, as Table 22-1 demonstrates.

Thus we are in a "distressing" situation that Ricardo complained about in his famous chapter "On Machinery" and which I call technological stalemate. To characterize this situation I quote from Fellner's article on the rate and direction of innovation:

Table 22-1. Types of Innovations, 1952–1973

Period	Type of Innovation		E/M
	Expansionary (E)	Modifying (M)	
1952–1954	29	36 ⎫	
1955–1959	8	33 ⎭	0.54
1960–1964	18	44	0.41
1965–1969	12	55	0.22
1970–1973	17	70	0.24
Total	84	238	

A high rate of innovation could not continue for very long if it became associated with a sufficiently pronounced maldistribution of the factor-saving effects.[1]

Needless to say, today we suffer from the labor-saving effects of a maldistributed technological change of one decade or two.

We are, I believe, living in an era of accelerated innovation, and such an era could become one of increasingly pronounced difficulties as a result of the overshooting of the labor-saving or of the capital-saving effect.[1]

If we look at the Eurodollar market, we see we are not suffering from a lack of capital. Therefore, structural instability is both the cause of current crises as well as a chance for change. Since structural readiness for major innovations has always been one characteristic of a stalemate in technology, we may expect rapid technological advances of the expansionary type to come in the long term. The current structural policy problem is how can we select such directions of innovation that could be brought about earlier than the market process would achieve them at its own pace? How can we be sure that these new lines of industry are going in a successful direction so that people can trust in progress and not fight it or withdraw from the opportunities it offers?

23 Summary

This chapter summarzes the most significant aspects from previous portions of the book. The ideas discussed earlier are brought to bear on important questions. First, what are the present context and future directions for technological innovation? What returns can be expected from industrial innovation? What the the manpower implications of technological innovations? What is the impact of technological innovation on international trade patterns? What are the human and social ramifications of technological change? What are the general trends regarding innovation in various countries? How important is the role of new-technology-based firms? What government policy options regarding innovation warrant further consideration? What are the effects of regulations on innovation? How can quasiexperiments be used to improve the management of technology? What is the overall climate for industrial innovation today and how can industry and government enhance innovation processes?

Present and Future Context for Innovation

There are unlikely to be any major technical barriers in manufacturing industries to hinder economic progress in the next decade. Problems of sustaining the rate and direction of technical change, however, could emerge because of low rates of investments or insufficient or inappropriate government investment in long-term research-and-development activities. In the low-tariff and industrially interdependent Western countries and Japan, uneven patterns of technical change will continue to pose problems. Uneven rates of technical change in one country could lead to pressures for protectionism or large-scale government intervention.

The rate and direction of technical change is likely to be different in the next decade. Innovation directed toward resource savings of all sorts will be more important, and different-patterns of consumer and government demands will emerge.

Returns from Industrial Innovation

It is increasingly important to understand the difference between social and private rates of return from investments in research and development and technological innovation. Until recently we had no evidence whether innovation mainly benefited the innovating firm, the users, or the recipients of technological change. We now have the beginning of empirical evidence that the major gains accrue to the users, which, if verified, has important policy implications. The initial findings are based on carefully conducted research on the returns from 17 innovations, and further studies are currently underway.

The social benefit of innovation is measured by considering the savings to users. For example, a product innovation in the primary metals industry resulted in a potential savings to manufacturers of household appliances. The private rate of return calculation involves subtracting all costs incurred by the innovator from the firms' revenues that were accrued from the innovation.

The innovations under investigation occurred in a wide variety of industries. Studies found that the social rates of return (returns to the users) from the investments in the innovations studied were very hgh. The median private rate of return (return to the innovating firm) from the investments in these innovations were much lower than the social rates of return to the users).

It is important to recognize the riskiness of research-and-development investments, evidenced by the enormous variation in the private rate of return among the 17 innovations studied. The results suggest that if the marginal social rate of return from investments in civilian technology is greater than the marginal social rate of return from other uses of relevant resources, this is evidence of an underinvestment in civilian technology. The results imply that there may be an underinvestment research and development and technological innovation in the United States, since the average social rate of return to the investing firm in the few innovations thus far studied in this way.

The Effects of Technological Change in Employment

Technological change has always been an important component affecting manpower requirements in industry. Most of the radical innovations in the textile machinery field, usually not considered high technology, have been associated with the presence of qualified engineers and sometimes scientists. The technological change in the textile machinery industry, for example, has significantly influenced the manpower component of textile production. The use of new technologies in textile production has forced textile firms to employ personnel with a greater range of skills.

Twenty textile companies that were successful in producing radical textile machinery innovations and 15 companies that produced only incremental innovations were studied. It was concluded that since radical innovations are important for the success of textile machinery companies, some way must be found to inject the necessary high level of technical expertise in the companies—particularly smaller companies—that currently do not possess the necessary expertise. It is important to consider techniques for high-technology inputs for traditional industries that still form the backbone of our economies.

Technological Innovation and International Trade Patterns

The role of technological innovation in explaining the level and composition of a country's exports has recently gained the forefront in international trade theory. Recent movements in the Organization for Economic Cooperation and Development (OECD) trade flows were examined to see if a technology factor is really identifiable and, if so, whether it conforms to generally held expectations.

The United States, the United Kingdom, Germany, France, and Japan account for 5% of OECD research-and-development expenditures. These countries may be characterized as research-rich and thus may be expected to have a comparative advantage in the export of technology-intensive products.

The technology factor does not distinguish itself in this measure of the trade characteristics of the OEDC countries with the excepton of the United States. The export patterns for the research-rich countries other than the United States are not notably different from those of research-poor countries. Technology-intensive products comprise 25–30% of total manufactures exports for each of the research-rich countries, as well as for the rest of the OECD.

However, 40% of United States exports of manufactures are technology-intensive products. Moreover, there is a high degree of correlation between the United States' share of total OEDC exports of a particular product and that product's research intensity.

In addition, even though technology-intensive products are commonly believed to be one of the most dynamic forces in world trade, the growth rate of these products is only slightly above that of non-technology-intensive products. Technology-intensive products are only very gradually accounting for an increasing share of OEDC exports of manufactures. On the whole, those countries that perform well in their technology-intensive exports also experience good growth in their nontechnology-intensive exports and vice vera. Even though there are some indications of technology working at the margin, the bulk of the evidence suggests that a country's exports, whether technology- or nontechnology-inetnsive, share factors other than technology in common that account for their relative competitive performance.

Thus technology is only one factor out of many influencing trade flows. Although attention has recently been focused on the transitory, monopolistic advantages of technological innovation, the primary influence of technology on international trade may well be indirect, through its contribution to overall productivity growth. As such, the measure of technology via research-and-development expenditures is not an adequate assessment of the overall importance of technological innovation to the exports of the developed countries.

Human and Social Aspects of Technological Change

The subject of human and social consequences of technological change was examined through a discussion of Sweden, a country with an extraordinarily high standard of living. Sweden has been on the forefront in its social policies directed toward humanizing work.

For example, consider process innovations at Volvo, where the management believed that people, not machines, are the real basis for the spectacular growth of industry during the twentieth century. Both cars and engines are assembled at the Volvo factories in working groups rather than assembly lines. Job rotation and decisions delegated beyond the norm are tradition in the Volvo factories.

The Volvo approach is not necessarily best for all workers. There are indications that psychosomatic complaints have risen in some of those employed in this participative way. This is speculatively associated with unwanted increased responsibility and decsion-making power.

Sweden has shown its willingnes to experiment with workplace innovations within its plants with significant human and social ramifications. The government has been a prime mover in encouraging these changes through a whole series of laws. The Swedish experience provides new options to be considered by government and industry from other countries.

The Environment for Industrial Innovation in the United States

Many argue that innovation has diminished in the United States in recent years and that the technical base for innovation is eroding, citing the National Science Board's *Science Indicators* as evidence. In addition, a large portion of industrial research and development is concentrated in only a few industries and a small number of companies within those industries. Industry funding in the United States for research and development is barely keeping up with inflation, and government funding for research and development is even less encouraging.

Furthermore, in addition to the United States' limiting incremental expenditures for research and developing, we must consider how the monies are being spent. Large expenditures are being made for research to comply with government rgulations. Companies also claim to be spending inordinately high sums to comply with regulatory statutes and practices, which in some cases may aid society but which in others may result in a massive misuse of technological and other resources. Many companies are now working on shorter-term research and development than in the past. With such uncertainty, United States companies are sacrificing long-term programs for programs of expediency aimed at short-term returns.

New Technology-Based Firms

Many innovations as well as sources of employment emanate from new-high-technology-based firms in the United States. However, do conditions also exist in Germany and England for such a situation?

The environmental factors of West Germany and the United Kingdom that concern the development of new-technology-based firms was examined by considering the two countries in terms of six factors: (1) their national research-and-development systems, (2) the role of venture capital, (3) taxation and forms of incorporation, (4) the patent system, (5) economic conditions, including investment and growth, and (6) behavioral and attitudinal factors.

The research included a review of the public literature plus approximately 100 interviews with experts in various subject areas and consultations with tax and patent experts in both England and Germany.

The study concluded that the number of new-technology-based firms in both England and Germany is low, and their performance has been unimpressive, particularly in comparison with the United States. In each of the two countries, there have been few successes, whether measured in terms of firm numbers, firm size, growth, or employment. The relatively poor showing of these new-technology-based firms in England and Germany is due to several factors.

By contrast, favorable factors in the United States include a large domestic market, the availability of private capital (although first-stage venture capital in the United States is far more difficult to obtain than it has been in the past), a fiscal framework that encourages the flow of private capital into a new ventures, the existence of an active market for the trading of shares in new ventures, and a prevailing attitude in the society that encourages entrepreneurship.

The Small, High-Technology Firms

This analysis, presented on the heels of the previously described study, cites some of the difficulties in the United States for small, high-technology firms. Starting a small, technology-based firm in the United States is still quite difficult, although apparently not as difficult as in England or Germany. It is extremely difficult for small, high-technology firms in the United States to do business directly with the Department of Defense as well as other government agencies. The magnitude of the paper work and increasing government requirements are major deterrents. As a practical matter, venture capital in the United States is hard to obtain. It may be a long time before our society starts to increase its level of financial risk to the point where venture capital becomes more readily available.

Several recommendations to encourage the formation of small high-technology firms are as follows: (1) establish a system for founders' stock for tax advantages, (2) recognize the role of corporate investors, (3) provide tax incentives for direct investment in small technical enterprises, (4) improve SEC rules regarding public offerings for small, high-technology firms, (5) review and establish various other incentive formats, and (6) increase the ease of perform-

Summary 251

ing under government contracts so that the valuable contributions from small innovative companies can again be made available to the country.

Government Policy and Innovation in England

The role played by research and development in individual firms has been widely believed to be essential for the maintenance of a technically progressve industrial sector. The idea of research and development as a national resource was accepted as a remedy for a wide range of industrial problems in Britain. When the results were disappointing, the government questioned the relationship between the exstence of research capability and the achievement of national goals. This has been true not only in the United Kingdom but in other countries as well.

An attempt was made in Britain between 1970 and 1977 to orient government-sponsored civilian research and development more in the direction of national needs. The implications of these reforms within the context of innovation in British industry are debatable. The organizational changes have been considerable over these years; and consquently, it is still too early to fully evaluate their effects.

However, some preliminary observation can be made about the likely impact of the changes on the concept of idustrial innovation in Britain. It was presupposed that the government knows what it wants and can articulate these wants and that industry can respond. This notion is based on an oversimplified view of the relationship that exists between a customer and a supplier in a traditional market-oriented business environment. Certainly the Department of Industry was aware of its role as a proxy customer for industry when it established the research requirement system. Skepticism exists as to the impact of government reforms on the innovative performance of British industry.

Government Policy and Innovation in Japan

Technological innovation, government policy, and motivated labor forces are the major factors that made possible Japan's rapid industrial expansion. The strategy has been to quickly follow up on innovations originally made elsewhere and add significant improvements to establish international competency.

The total expenditure for research and development in Japan has increased at an annual rate of approximately 20% over the past 10 years. Japan ranks fourth among the major nations in total research-and-development expenditures, with the United States first, the U.S.S.R. second, and West Germany, third. Unlike Germany and the United States, Japan's government funding for research and development is only 26.5% of the total research and development as opposed to 50% in the United States, Germany, and most other OECD countries. Japan's successful growth is obviously not a function of government funding for research and development, but rather the systematic cooperation and delineation of responsibilities between government and industry.

However, Japan will face several significant problems in the future. As the innovation systems tend to become larger and more complex, the period required for their completion tends to be longer. On the other hand, it is almost certain that the accelerated and unpredictable changes in social and technological environment will occur at an increasing rate. Japan can depend to some extent on conventional methods of forecasting either by extrapolation or through consensus. However, the extrapolation approach cannot cope with discontinuous changes, and a consensus approach generally provides a qualitative expression of common sense. Japan, therefore, is searching to find better methods of system synthesis by which a system or its components can adaptably respond to the changes in the social and technological environment.

The basic cooperation between government and industry is likely to continue as a major reason for Japan's economic success. Japanese industry does not submit itself entirely to the unseen hand of the market but accepts government guidance. Concomitantly, the government does not stifle economic growth, as it does in some countries, by overplanning and overcontrolling the entire economy. In a broader sense, the cooperative relationship between government and industry in Japan is a complex phenomenon that is not well understood in Western countries.

Government Policy and Innovation in the United States

There is persuasive empirical evidence that research and development and technological innovation have had a positive effect on economic growth and productivity increases in the United States. However, economic returns from investments in research and development in the United States show a wide variance.

Summary

There is general agreement on why the market mechanism by itself is likely to lead to underinvestment in research and development from society's point of view. An investment in research and development involves a large element of risk, the more so as one approaches the basic research end of the spectrum. Studies have shown a great variability in industries' rates of return obtained from innovation, and the rates of return to society were about twice as high as the private rates of return to the firm itself.

Even though investigations have shown imperfections in the market mechanism from the social-returns point of view, this does not necessarily mean that government action is warranted. It is necessary to demonstrate that private returns are insufficient to call forth adequate investment in innovative activities and that proposed government policies and actions are cost effective.

No consensus exists as to whether health, safety or environmental regulation has been beneficial or detrimental to technological innovation. There are examples of both kinds of results and good reasons to believe that we should not expect general conclusions because of the different forms or regulations and their varying effects.

Several suggested options for public policy and practice regarding civilian innovation in the United States are: (1) government policy and practice regarding civilian-section research and development and technological innovation should be made more consistent with government economic and social policy; (2) government policy and practice should be more consistent over time; (3) government policy and practice should reinforce and help perfect private market forces rather than substitute for them.

Government Regulations and Innovation

There has been a continual increase in the amount of performance regulations in the United States, and the rate of increase is continuing to rise. Some companies claim that they expend more than 45% of their research-and-development funds to comply with federal regulations. In terms of decreased innovations, it has been argued that the drug industry in the United States has been turning out fewer and fewer innovations, in part because of the high requirements set by the Food and Drug Act and its administration.

However, the benefits from regulations are equally dramatic. The decrease in automobile deaths in the United States can be attributed both to regulated speed as well as safety features required on United

States vehicles and roads. Equally dramatic is the decrease in deaths due to aspirin poisoning, which have decreased from more than 8000 aspirin poisonings among children in 1972 (before legislation forced manufacturers to change the cap) to less than 5000 in 1974 (two years after enactment of the regulation).

Increased understanding between industry and government on the issue of regulation is necessary. Government often regulates with little input from industry personnel who are close to particular products. Societal impact analyses conducted prior to regulations may be useful as well as ripple effect studies to carefully weigh the regulations' total effects. In addition, a periodic assessment of government regulations and elimination of outdated laws and conflicting laws would be beneficial, as would research programs and experiments to decrease uncertainties. There is great value in a cost–benefit conceptual approach to better understand the full ramifications of the effects of regulation upon innovation.

Improving the Management of Technology

Governments could and should make more systematic use of data that are collected from research-and-development projects. The objective is for governments to use these data to learn ways to improve the management of research and development. Governments should set up quasiexperiments to improve the management of research and development.

Small businesses and even medium-sized and large companies invest substantial funds and have major problems in regard to research and development. Similarly, governments spend significant funds to support research and development, and they should be able to do this more effectively. Experiments could be used to learn more about how to manage research and development. True experimental designs are characterized by a randomized assignment of units for experimental treatments. The support of research-and-development projects is certainly not designed as a random process, and the experiments would, therefore, be quasi rather than true.

Several quasiexperimental designs can be used to gather data with the objective of improving the research-and-development process. The successful application of quasiexperiments would be of importance both to the sponsoring organization (the government) as well as to the recipients of the research funds (industry).

Closer Integration of Science Policy and Economic Policy

The separation between science policy and economic policy in the United States is a serious obstacle to the design of sensible policies. For example, the objective of holding down housing costs is unlikely to be accomplished without significant technical advances in housing. Furthermore in the automotive industry the lack of integration between technology and economics has been largely responsible for inaction in the past.

The re-establishment of a science policy office in the Executive Office in the United States provides an opportunity to overcome some of the difficulties. However, the problem is larger than merely establishing a "presence" for science in the Executive Office. The objective is to mesh a perspective on what can evolve technologically with a sophisticated assessment of the economic implications and appropriate role of the federal government in facilitating the evolution of that technology.

A change in perceptions in all three of the key United States' agencies is required. The Council of Economic Advisors has evolved over the years into an organization much more concerned with microeconomic policies than it used to be. The new Office of Science and Technological Policy must see science policy as an instrument that needs to be tailored to fit an overall policy design. Finally, the Office of Management and Budget (OMB) is in the best position to require that the design of an overall policy specifically involve a research-and-development component and that science policy be linked to an overall policy design.

Innovation Strategies for Government and Industry

Innovations are of prime importance for Europe (as well as the rest of the world), since without them Europe's economic position would deteriorate, to say nothing of the health and welfare of the people. Innovations involves a whole series of both technical and productive and commercial factors. Some industries and countries in Europe are still largely anchored to the past and show fear of the new. Europe, in some areas, has an attitude of defense and self-protection rather than an aggressive risk-taking spirit. Europe has a great tradition for scientific research, although the spirit of liveliness and the entrepreneurial thrust have often been missing, in contrast to the United States, where the spirit continues to be a true characteristic of the country.

The legislative and the educational systems in the United States are structured to factor and enhance the value of the entrepreneurial drive. In the United States, advanced research has been successful in areas such as aerospace, computers, automation, and nonconventional materials. This, in turn, encourages innovation in other sections of the economy. However, the strongest contributing factor to the success of the United States in innovation has been its response to market-pull rather than technology-push.

To achieve a stable position of competitiveness, European counries must derive a technological policy and be ready to adjust the policy to the rapid changes of the world's economic and sociopolitical environment. It is not clear that a European technological policy is feasibile, but a desirable goal is to integrate the individual countries technological policies. In European countries (as in many other countries) it would be a basically poor policy to protect weak economic sectors. There should be an emphasis on decentralization. There should be a better regional management.

Some recommendations include the following:

1. Stimulate industrial competition as a main pressure for technological innovation
2. Ensure a creditable reward for innovations through the tax and patent systems
3. Regulate codes and standards so that they take into account the social costs and benefits of the innovative process
4. Have active regional and manpower policies that deal with the changes in skill patterns brought about by technological change
5. Use government procurement, where appropriate, to upgrade the technical level of industry
6. Encourage the mobility of scientists and engineers, especially in and out of government laboratories
7. Identify policy measures to encourage innovation

Aligning Industry Planning and Government Policy

The oldest government measures with direct implications for industrial research and development are undoubtedly the patent laws.

Summary

These laws provide for the protection of the individual inventor, giving him the possibility to reap the financial benefits from his own invention. Another classical field in which government policy has consequences for research-and-development activity in industry is that of scientific and technical education.

In Europe and the United States the governments' role in research and development is often based on national defense. In addition, government measures for safeguarding the health and safety of individual citizens affects industrial research and development. An example is the scrutiny of new pharmaceuticals, which is much more severe today than it was several year ago.

In general, the management of a private enterprise will have the goal of minimizing direct government influence on its policy, its strategy, and its operations. The timely interpretation of new directions as a result of government legislation is of more importance to a company than financial support for its research and development.

The vastly increased complexity of today's society calls for regulatory power in certain areas. Private industry must realize how poorly government is equipped to deal with certain industrial questions requiring highly specialized expertise. The most realistic position is to accept the increasing interaction between government and industry in the field of research and development and devote systematic attention to these interactions and be prepared to use the situation in the most fruitful way for society and for enterprises.

24 Notes

Chapter 3

1. E. Mansfield, I. Rapoport, A. Romeo, S. Wagner, and G. Beardsley, "Social and Private Rates of Return from Industrial Innovations," *Quarterly Journal of Economics*, Vol. XVI, No. 2 (May 1977), pp. 221-240; E. Mansfield, J. Rapoport, A. Romeo, E. Villani, S. Wagner, and F. Husic, *The Production and Application of New Industrial Technology* (New York: Norton, 1977).

2. Z. Griliches, "Research Costs and Social Returns: Hybrid Corn and Related Innovations," *Journal of Political Economy*, Vol. LXVI, No. 5 (October 1958), pp. 419-431.

3. K. Arrow, "Economic Welfare and the Allocation of Resources for Invention," *The Rate and Direction of Inventive Activity* (Princeton, N.J.: Princeton Univ. Press, 1962); R. Matthews, "The Contribution of Science and Technology to Economic Development," in *Science and Technology in Economic Growth*, B. Williams (Ed.) (New York: Macmillan, 1973).

4. See National Science Foundation, *Research and Development and Economic Growth/Productivity* (Washington, D.C.: Government Printing Office, 1971).

5. E. Mansfield, "Federal Support of R and D Activities in the Private Sector," *Priorities and Efficiency in Federal Research and Development* (Washington, D.C.: Joint Economic Committee of Congress, October 29, 1976); E. Mansfield, "Role of Technological Change in U.S. Economic Growth, unpublished paper presented to Fifth World Congress of the Internationadl Economic Association, Tokyo, 1977, to be published by Macmillan in the conference proceedings.

Chapter 4

1. R. Rothwell, *Innovation in Textile Machinery: Some Significant Factors in Success and Failure*, Science Policy Research Unit Occasional Paper Series No. 2 (June 1976).

2. R. Rothwell, "Technological Innovation in Texitle Machinery: The Role of Radical and Incremental Change," *Textile Institute and Industry*, Vol. 14, No. 11 (November 1976), pp. 330-336.

3. H. Catling and R. Rothwell, "Automation in Textile Machinery," *Research Policy*, Vol. 6 (1977), pp. 164-176.

4. E. Cummins, Conference Report: "Combined English Mills: A Case Study in Rationalization," *Viyella International* (Autumn 1966), pp. 4-10.

Chapter 5

1. For a detailed explanation of the methodology, see R. Kelly, *Alternative Measures of Technology-Intensive Trade*, OER/ER-17 (Washington, D.C.: Department of Commerce, 1976).

2. In the course of analysis, the use of two other factors commonly used as technology proxies was rejected: 1) the "concentration of scientists and engineers engaged in R and D" was found to be redundant with research-and-development intensity (these two measures have a 96% correlation); and 2) "the relative hourly wage earnings of production workers" was rejected as being more reflective of such factors as industry concentration, union strength, and so on, than of technology-intensity.

3. The most notable exceptions to the U.S. pattern were found to be iron and steel products and ships, which receive relatively greater research-and-development emphasis in some other OECD countries than in the United States. Even in these instances, however, the differences were not great enough to warrant their inclusion in the technology-intensive category.

4. Two points should be noted about the data presented in this paper. First, no corrections were made to "the rest of the OECD" for the inclusion of Finland in 1971 and Australia and New Zealand in 1974. These three countries accounted for 5.2 and 1.8% of the total manufactures of the "rest of the OECD" and total OECD, respectively, in 1974; hence, these three countries are primarily responsible for the larger share of exports of the "rest of the OECD" relatively to the five "research-rich" countries in 1974 compared to 1968.

Second, because of supply/demand imbalances, OECD exports of iron and steel mill products rose 80% in 1973–1974, compared with a compound annual rate of 21% from 1968–1973. This abberation resulted in iron and steel products being the fastest growing OECD export commodity for the period. Consequently, 1974 iron and steel exports were corrected (reduced) by allowing the iron and steel exports of each of the major OECD countries in 1974 to increase by their compound annual rate of growth in the previous 5-year period.

5. The OECD export trade for each of the five "research-rich" countries and the "rest of the OECD" in each of the product classes was deflated on the basis of the United Nation's dollar unit value indices for total manufacturing, published in the *Monthly Bulletin of Statistics*. Although this procedure obviously introduces an error into the deflated values for "technology-intensive" and "nontechnology-intensive" trade, it yielded superior results to a discussion based entirely on current dollar values.

Chapter 7

1. E. Mansfield, "Contribution of R and D to Economic Growth in the United States," *Science*, Vol. 175 (February 4, 1972), pp. 477–486.

2. Arthur D. Little, "New Technology-Based Firms in the United Kingdom and the Federal Republic of Germany," (London: Wilton House Publications Ltd., 1977).

3. "The Breakdown of U.S. Innovation," *Business Week*, (February 16, 1976), pp. 56–68.

4. National Science Board, National Science Foundation, "Science

Indicators 1974," (Washington, D.C.: U.S. Government Printing Office, December 1975).

5. Analysis by P. M. Boffey, "Science Indicators New Report Finds U.S. Performance Weakening," *Science*, Vol. 191 (March 12, 1976), pp. 1031–1033.

6. B. Ancker-Johnson, "National Science and Technology Policy-Current Policies and Options for the Future," *Research Management*, Vol. 20 (January 1977), pp. 7–12.

7. W. H. Fisher et al., "Probable Levels of R and D Expenditures in 1977, Forecasts and Analysis," (Columbus, Ohio: Battelle Columbus Laboratories, December 1966).

8. "Where Private Industry Puts Its Research Money," *Business Week* (June 28, 1976), pp. 62–84.

9. "How GM Manages Its Billion-Dollar R and D Program," *Business Week* (June 28, 1976), pp. 54–58.

10. "Old Products Remain Top R and D Priority," *Chemical Week* (May 25, 1977), pp. 39–40.

11. "Research Management Trends: 1974–1977," *Research Management*, Vol. 20 (January 1977), p. 4.

12. "Old Products Remain Top R and D Priority," pp. 39–40.

13. P. F. Chenea, "Impact of Regulations on R and D; The Costs and Effects of Regulations,' *Research Management*, Vol. 20 (March 1977), pp. 22–26.

14. "Detroit's Response to the Energy Problem," *Business Week* (May 23, 1977), p. 100.

15. J. D. Lewis, "National Science and Technology Policy; Its Impact on Technological Change," *Research Management*, Vol. 20 (January 1977), pp. 13–16; D. Gillette, "Impact of Regulations of R an D; How Regulations Encourage and Discourage Innovation," *Research Management*, Vol. 20 (March 1977), pp. 18–21.

16. "R&D," *Astronauts and Aeronautics*, Vol. 15 (March 1977), p. 28.

17. "Old Products Remain Top R and D Priority," pp. 39–40.

18. *Research Management*, Vol. 20 (January 1977), pp. 3–4.

19. *Research Management*, Vol. 20 (March 1977), pp. 3–4.

20. "Detroit's Response to the Energy Problem," p. 100.

21. "Science at the Bicentennial—A Report from the Research Community," Report of the National Science Board/1976, pp. 28–29.

22. I. S. Shapiro, "Why Not Zero Based Regulation," an address to the Atlanta, Georgia, Rotary Club, May 23, 1977.

23. M. L. Weidenbaum, "A Fundamental Reform of Government Regulation" (St. Louis, Mo.: Center for the Study of American Business, Washington University).

24. American Society of Association Executives, "Technological Objectives and Federal Policy," A Report to the Experimental R and D Incentives Program (Washington, D.C.: National Science Foundation, July 1976), 2 volumes.

25. F. R. Bradbury, "Constraints to Innovation," *Chemtech* (January 1977), pp. 23–24.

26. "Technology, Productivity and Economic Growth," National Science Foundation, Advanced Productivity Research and Technology Divi-

sion, Mosaic, September/October 1976. Reprinted in UNIT, November 1976.

Chapter 9

1. At the same time, one may also hypothesize that social (systems) innovations will play a more dominant role in the future.
2. This is the definition used by the R & D Assessment Program of the National Science Foundation. See L. L. Ledermann: "Technological Innovation and Federal Government Policy—Research and Analysis of the Office of National R&D Assessment," in *Innovation, Economic Change and Technology Policies*, Karl A. Stroetmann (Ed.) (Basel: Birkhauser, 1977), p. 188.
3. See W. Langen, "Sekroraler Strukturwandel und Unternehmensgrosse," in *Unternehmensgrössenstatistik* (Bonn: Institut fur Mittelstandsforschung, Studienreihe des BMWi Nr. 16, December 1976), pp. 76–88.
4. G. Fels and E.-J. Horn, "Kleine und mittlerer Unternehmen im Prozess des weltwirtschaftlichen Strukturwandels," in *Die gesamtwirtschaftliche Funktion kleiner und mittlerer Unternehmen*, K. H. Oppenlander (Ed.), (München: IFO, 1976), pp. 185–206.
5. *Fifth Report of the Federal Government on Research* (Bonn: Federal Ministry for Research and Technology, 1976), p. 11.
6. "The Smart Machine Revolution," *Business Week* (July 5, 1976), p. 44.
7. See, for example, Charleswater Assn., *The Impact on Small Business Concerns of Government Regulations that Force Technical Change*, Report prepared for Small Business Administration and Experimental Technology Incentives Program (Boston, September 1975).
8. See, for example, the literature reviewed in M. I. Kamien and N. L. Schwartz, "Market Structure and Innovation: A Survey," *Journal of Economic Literature*, Vol. 13, No. 1 (March 1975), pp. 1–37.
9. Often these can be traced to limited opportunities to make use of economies of scale or to diversify risk, to the absence of synergetic effects, and so on.
10. Infratest-Industria, *Informationsbedarf und Informationsversorgung der mittelständischen Indusrtie in Bayern*, report (München: Bayerisches Staatsministerium fur Wirtschaft und Verkehr, July 1975).
11. On this see H. Echterhoff-Severitt, "Expenditures for Research and Development in Business Enterprises of the Federal Republic of Germany in 1971 and 1973," in *Innovation, Economic Change and Technology Policies*, K. A. Stroetmann (Ed.) (Basel: Birkhauser, 1977), p. 74.
12. H. Echterhoff-Severitt et al., *Forschung und Entwicklung in der Wirtschaft 1973* (Essen: Stifterverband fur die deutsche Wissenschaft, 1977).
13. Ibid.
14. Ch. Rothlingshofer, *Die Vergabe von Forschungsund Entwicklungsauftragen in der BRD* (Berlin: Duncker & Humblot, 1972).
15. K. Orths, "Von der Gemeinschaftsforschung zur Innovation in der Giebereiindustrie," *Giebereiforschung*, Vol. 27, No. 2 (1975), pp. 52–53.
16. See R. M. Colton and G. G. Udell, "The National Science Founda-

tion's Innovation Centers—An Experiment in Training Potential Entrepreneurs and Innovators," *Journal of Small Business Management*, Vol. 14, No. 2 (April 1976), pp. 11–20 and "University Business Development Center," SBA-Paper (March 1977).

Chapter 10

1. See pp. 80.
2. A. C. Cooper and A. V. Bruno, "Success Among High-Technology Firms," *Business Horizons* (April 1977), pp. 16–22.
3. J. S. Gansler, "Let's Change the Way the Pentagon Does Business," *Harvard Business Review* (May–June 1977), p. 109.
4. P. R. Payne, "A Comparison Between 'Small' and 'Large' R&D Companies," ERDA/AASRC Conference & Workshop, Washington, D.C., March 24–25, 1975.
5. See pp. 159.
6. A venture capitalist would say "there's no market window for floating the company." Morse[10] gives the following numbers for new small company public issues:

1972	104 issues
1973	19 issues
1974	4 issues
1975	0 issues

7. Plato pointed out a long time ago that there's a basic self-destruct mechanism built into democracy, in that its freedom produces demogogues who can often sway a majority of the voters into unwise choices. Bearing in mind his views of his contemporary theater, I should love to have heard his views on our modern media!
8. J. Rabinow, Testimony before the Subcommittee on Domestic and International Scientific Planning and Analysis of the Committee on Science and Technology of the Congress of the United States, April 29, 1976.
9. C. N. Parkinson, *The Law and the Profits* (New York: Houghton-Mifflin, 1960).
10. R. S. Morse, "Innovative Technology: What Is Its Impact on U.S. Economy?" *Professional Engineer* (August 1976), pp. 14–26.
11. R. A. Charpie, Chmn., "Technological Innovation: Its Environment and Management," Panel on Invention and Innovation, United States Department of Commerce, January 1976.
12. Agriculture-Enviromental and Consumer Protection Appropriations for 1975—Part 7, Investigative Report on the Utilization of Government Laboratories, Hearings before a Subcommittee of the Committee on Appropriations, House of Representatives, 93rd Congress, 2nd Session.

Chapter 11

1. See above, pp. 80.
2. American Council on Independent Laboratories (ACIL), American Association of Small Research Companies (AASRC), Consulting Engineers Council (CEC), National Council of Professional Services Firms (NCPSF) all publish directories.

Notes   263

3. J. McFarlane, Chmn., Laboratory Management Committee, ACIL, private communication.
4. P. R. Payne, "Case Histories of Small R&D Company Accomplishments," Proceedings of Conference—"Opportunities at ERDA for Small R&D Companies," Conf-760360. "ACIL Bulletin," Vol. 16, No. 2 (Fall 1976), American Council of Independent Laboratories, Washington, D.C. Jacob Rabinow, Testimony before the Subcommittee on Domestic and International Planning and Analysis of the Committee on Science and Technology of the Congress of the United States on April 29, 1976.
5. A. Zaleznik, "Managers and Leaders: Are They Different?" *Harvard Business Review* (May–June 1977), p. 67.
6. Our company and thousands of others like us were forced to cancel perfectly sound pension plans as a result of the red tape requirements of the Pension Reform Act of 1975.
7. Investigative Report on Utilization of Federal Laboratories, Agriculture-Environmental and Consumer Protection Appropriations for 1975—Part 7.
8. Memorandum Concerning Unfair Competition by Universities and other "Non-Profit" Organizations in the Research and Testing Business, Eastern Division ACIL, March 30, 1977.
9. M. Maher, Chairman, Committee E-36, ASTM, "Standard Recommended Practice for Generic Criteria for Use in the Evaluation of Testing and/or Inspection Agencies, American Society for Testing and Materials" (ASTM) Standard #-548. Manual of Practise, Quality Control System Requirements for a Testing and Inspection Laboratory, American Council of Independent Laboratories, Washington, Pub. #LA(62)1-76.

Chapter 12

1. C. Freeman, *The Economics of Innovation* (New York: Penguin Books, 1974), especially Chapter 8.
2. See, for example, *A Framework for Government Research and Development.* Cmnd. 4814, HMSO, 1971, and K. Pavitt and W. Walker, "Government Policies Toward Industrial Innovation: A Review," *Research Policy*, Vol. V, No. 1 (January 1976), also P. Aucoin and R. French, "Knowledge, Power and Public Policy," Science Council of Canada, *Report 31* (November 1974).
3. *A Framework for Government Research and Development*, Cmnd. 4814, HMSO, 1971, p. 3.
4. M. Gibbons and P. J. Gummett, *Redeployment in Government Research Establishments*, a report submitted to the Science Council of Canada, August 1975.
5. House of Commons 375 (1971–1972), pp. 329, 459–460.
6. Ibid., p. 460, question 18.
7. Ibid., p. 329, and passim.
8. *Framework for Government Research and Development*, Cmnd. 5046, HMSO, 1972.
9. Department of Trade and Industry, *The Non-Reactor Research and Development Activities of the Atomic Energy Authority and the Industrial Research Establishments of the Department of Trade and Industry*, Cmnd. 5176, HMSO, 1972, p. 9.

10. House of Commons, 375, (1971–1972), p. 465, question 49.
11. M. Gibbons and P. J. Gummett, "Recent Changes in the Administration of Government Research and Development in Britain," *Public Administration* (Autumn 1976), pp. 247–266.
12. For further details, see P. J. Gummett and Michael Gibbons, "Government Research for Industry: Recent British Developments," *Research Policy*, in press.
13. Department of Industry, *Interim Strategies of the Research and Development Requirements Boards* (London: Department of Industry, 1975).
14. See Gummett and Gibbons, op. cit.
15. Department of Industry, *Reports on the Research Requirements Boards, 1973* (London: Department of Industry, 1974), p. 2.
16. Ibid.
17. See *Research and Development Requirements Boards, Reports 1974–75* (London: Department of Industry, 1975), p. 11.
18. Department of Industry, *Reports of the Research Requirements Boards, 1973*, p. 16.
19. Ibid., p. 34.
20. Department of Industry, *Interim Strategies of the Research and Development Requirements Boards*, p. 5.
21. Ibid., p. 7.
22. See Ibid., pp. 9–10, for full details of the report from the Chief Scientist's Board.
23. Ibid., pp. 11–13.
24. Department of Industry, *Reports of the Research Requirements Boards, 1973*, pp. 54–55.
25. See *Research and Development Requirements Boards, Reports 1974–75*, p. 35 and *Reports 1975–76*, p. 38, respectively; and *Ship and Marine Technology Requirements Boards, Report, 1975* (London: Department of Industry, 1976).
26. Department of Industry, *Interim Strategies of the Research and Development Requirements Boards*, p. 13.
27. House of Commons 375, p. 458, questions 3 and 4.
28. House of Commons 375, p. 465, question 49.
29. *Research and Development Requirements Boards, Reports 1975–76*, p. 7.
30. See, for example, R. R. Nelson and S. G. Winter, "Neoclassical or Evolutionary Theories of Economic Growth: Critique and Prospectus," *Economic Journal* (December 1974), pp. 887–889.
31. W. Nordhaus and J. Tobin, "Is Growth Obsolete," in *Economic Research: Retrospect and Prospect, Economic Growth*, R. Gordon, (Ed.), (New York: National Bureau of Economic Research, 1972), p. 2; also quoted in Nelson and Winter, "Theories of Economic Growth," p. 889.
32. Nelson and Winter, "Theories of Economic Growth," passim.
33. See, M. Gibbons and R. Johnston, "The Roles of Science in Technological Innovation," *Research Policy*, Vol. 3 (1974), p. 220.
34. J. Langrish, M. Gibbons, W. G. Evans, and F. R. Jevons, *Wealth from Knowledge* (New York: Macmillan, 1972), p. 78.
35. Ibid., pp. 66ff. See also, "Success and Failure in Industrial Innova-

tion," in *The Economics of Industrial Innovation*, C. Freeman (Ed.) (New York: Penguin Books, 1974), p. 161ff.

36. J. Langrish et al., *Wealth from Knowledge*, p. 42; and W. Gruber and D. Marquis (eds.), *Factors in the Transfer of Technology* (Boston: M.I.T. Press, 1969), pp. 11–23.

37. Memorandum of the Industrial Strategy Staff Group, *Industrial Strategy: Analysis of Sector Reports* (London: National Economic Development Office, 1976), passim.

Chapter 14

1. "Technological Innovation and Federal Government Podlicy—Research and Analysis of the Office of National R&D Assessment," NSF 76-9 (Washington, D.C.: National Science Foundation, January 1976).

2. "Research and Development and Economic Growth/Productivity—Papers and Proceedings of a Colloquium," NSF 72-303 (Washington, D.C.: U.S. Government Printing Office, December 1971), Stock No. 3800-0117.

3. E. Mansfield, J. Rapoport, et al., "Social and Private Rates of Return from Industrial Innovations," University of Pennsylvania, September, 1975 (available from NTIS,* Vol. 1, Analytical Report: No. PB-254 083/AS, $4.00; Vol. II, Case Studies: No. PB-254 084/AS, $4.50).

4. A. Bean, D. Schiffel, and M. Mogee, "The Venture Capital Market and Technological Innovation," *Research Policy*, Vol. IV, No. 4 (October 1975), pp. 380–408. Also, "Technology Startup—Venture Money Market Easing," *Industrial Research* (March 1977), p. 36.

5. R. Gilpin, "Technology, Economic Growth, and International Competitiveness," A Report prepared for the Subcommittee on Economic Growth of the Joint Economic Committee, Congress of the United States (Washington, D.C.: U.S. Government Printing Office, 1975), Stock No. 052-070-0300406.

6. See, for example, "U.S. Technology Policy—A Draft Study," Office of the Assistant Secretary for Science and Technology, U.S. Department of Commerce, March 31, 1977 (available from NTIS,* No. PB-263 806/AS, $9.50).

7. Gellman Research Associates, Inc., "Economic Regulation an Technological Innovation: A Cross-National Literature Survey and Analysis" (January 1974) (available from NTIS,* Vol. I, Summary and Analysis: No. PB-233 085/AS, $6.00; Vol. II, Part 1, Abstracts: No. PB-243 314/AS, $25.25; Vol. II, Part 2, Abstracts plus Negative Reports: No. PB-243 315/AS, $15.25).

8. R. C. Noll, "Government Policies and Technological Innovation," California Institute of Technology (available from NTIS,* Vol. I, Project Summary: No. PB-244 572/AS, $3.75; Vol. II, State-of-the-Art Surveys: No. PB-244 572/AS, $7.50; Vol. III, Research and Policy Studies: No. PB-244 573/AS, $6.25; Vol. IV, Abstracts: No. PB-244 574/AS, $12.25; Vol. V, Bibliography: No. PB-244 575/AS, $7.50; complete set Vols. I through V:

* National Technical Information Service, U.S. Department of Commerce, 5285 Port Royal Road, Springfield, Va. 22151. A microfiche copy of each document is available from NTIS for $2.35.

$42.00). C. Hill, "A State-of-the-Art Review of the Effects of Regulation on Technological Innovation in the Chemical and Allied Products Industry," Washington University, (available from NTIS,* Vol. I, Executive Summary: No. PB-243 727/AS, $3.25; Vol. II, The State of the Art: No. PB-243 728/AS, $7.25; Vol. III, Abstracts and Literature List: No. PB-243 729/AS, $7.50; 3-volume set: No. PB-243 726/AS, $16.00).

9. C. Trozzo and C. Kitti, *Effect of Patent and Antitrust Laws, Regulations, and Practices on Innovation* (Arlington, Va.: Institute for Defense Analyses, 1976) (Available from NTIS,* Vol. I, State of the Art: No. PB-252 860/AS, $10.00; Vol. II, Executive Summary: No. PB-252 861/AS, $3.50; Vol. III, Abstracts: No. PB-252 862/AS, $9.25; 3-volume set: No. PB-252 859, $20.00).

10. M. Kamien and N. Schwartz, "Market Structure and Innovation: A Survey," Northwestern University (June 1974) (available from NTIS,* Survey Summary, No. PB-235 649/AS, $3.00; Survey: No. PB-235 585/AS, $4.00; Survey Supplement-Precis/Annotated Bibliography: No. PB-235 648/AS, $5.50).

11. *The Effects of International Technology Transfers on U.S. Economy*, NSF 74-21 (Washington, D.C.: U.S. Government Printing Office, 1974), Stock No. 3800-00181.

Chapter 15

1. A. H. Rubenstein, "Models of the Innovation Process and Criteria for Evaluating Experiments," POMRAD, Department of Industrial Engineering and Management Sciences, Northwestern University. July 1972. A. H. Rubenstein, C. F. Douds, and T. W. Schlie, "Experimentation in the R&D & Technological Innovation Process for the Experimental R&D Incentives Program," *Technological Innovation*, Denver Research Institute (Boulder, Colorado: West View Press, June 1977), in press.

2. A. H. Rubenstein and J. Ettlie, "Final Report—Barriers to Technological Innovation Among Suppliers to the Automotive Industry: An Exploratory Study" (Evanston, Ill.: AHR Associates, March 1977).

3. A. P. Hurter, A. H. Rubenstein, J. Garvey, J. Martinich, and Mark Gergman, "Market Penetration by New Innovations: The Technological Literature," Department of Industrial Engineering and Management Sciences, Northwestern University, December 1976.

4. A. H. Rubenstein, A. K. Chakrabarti, and R. D. O'Keefe, "Final Technical Report on Field Studies of the Technological Innovation Process," in *Progress in Assessing Technological Innovation, 1974, Summary Reports of the National R&D Assessment Program*, (Ed.), (Westport, Conn.: Technomic Publishing Company, 1975), pp. 44–49.

5. A. H. Rubenstein and C. W. N. Thompson, "Preliminary Ideas on an Experiment to Test the Effects of Exclusive/Non-Exclusive Licensing," POMRAD, Department of Industrial Engineering and Management Sciences, Northwestern University, March 1975, a report to the Denver Research Institute.

6. A. H. Rubenstein and M. Radnor, "A Model of the Responses of Industrial Firms to Federal Procurement Incentives," a report to the National Bureau of Standards, June 1975.

7. A. H. Rubenstein, C. F. Douds, H. Geschka, T. Kawase, J. P. Miller,

R. St. Paul, and D. Watkins, "Management Perceptions of Government Incentives to Technological Innovations in England, France, West Germany, and Japan," *Research Policy*, Vol. 6, No. 4 (October 1977), pp. 324-357.

8. B. Peters-Koehler, "Public Support of Industrial Research and Development in the Federal Republic of Germany," POMRAD, Department of Industrial Engineering and Management Sciences, Northwestern University, February, 1973. R and D Research Unit, Manchester Business School, University of Manchester, Manchester, England, "Financing Development Projects: Method Used by the National Research Development Corporation," June 1973. R. St. Paul, "International Study of Innovation Incentives and Barriers, Centre D'Etudes Economiques D'Enterprises, Paris, France, and Department of Industrial Engineering and Management Sciences, Northwestern University, August 1974. H. Schwerdtner and H. Geschka, *Government Innovation Incentives and Their Impact on Innovative Behavior in Industry in the Federal Republic of Germany*, Batelle Institute, Frankfurt, Germany, and Department of Industrial Engineering and Management Sciences, Northwestern University, October 1974. P. Burke-Smith, A. Pearson, and D. Watkins, "International Study of Perceptions of Innovation Incentives and Barriers: Report on United Kingdom Participation in an International Comparative Study," R&D Research Unit, Manchester Business School, University of Manchester, Manchester, England, January 1975. T. Kawase and A. H. Rubenstein, "Reactions of Japanese Industrial Managers to Government Incentives to Innovation—An Empirical Study," *Transactions on Engineering Management* (August 1977), in press.

9. Rubenstein and Ettlie, "Final Report—Barriers to Technological Innovation Among Suppliers to the Automotive Industry."

Chapter 16

1. A. Gerstenfeld, "Technological Forecasting,' in *Methods and Techniques of Business Forecasting*, Robert A. Kavesh (Ed.) (Englewood Cliffs, N.J.: Prentice-Hall, 1974), pp. 222-238.

2. A. Gerstenfeld, "Interdependence and Innovation," *OMEGA: The International Journal of Management Science*, Vol. 5, No. 1 (1977), pp. 35-42.

3. A. Gerstenfeld, "A Study of Successful Projects, Unsuccessful Projects, and Projects in Process in West Germany," *IEEE Transaction on Engineering Management*, Vol. EM-23, No. 3 (August 1976), pp. 116-123.

4. A. Gerstenfeld, *Innovation: A Study of Technological Policy* (Washington, D.C.: University Press of America, 1976).

5. A. Gerstenfeld and L. Wortzel, "Strategies for Innovation in Developing Countries," *Sloan Management Review*, Vol. 19, No. 1 (Fall 1977), pp. 57-68.

6. See L. Lederman, "Federal Policies and Practices Related to R&D/Innovation," *Research Management*, Vol. XXI, No. 3 (May 1978), pp. 18-20.

7. P. F. Chenea, "The Costs and Effects of Regulations," *Research Management*, Vol. XX, No. 2 (March 1977), pp. 22-26.

8. H. E. Thayer, "Business in an Era of Legislation and Regulation," *Chemistry and Industry*, Vol. 19 (March 1977).

9. *Chemical and Engineering News* (March 14, 1977).
10. M. Weidenbaum, *Business, Government, and the Public* (Englewood Cliffs, N.J.: Prentice-Hall, 1977), pp. 16–18.
11. *Business Week* (February 21, 1977), pp. 80–84.
12. N. Ashford and G. Heaton, "Government Regulation: The Impact on Technological Innovation," *Professional Engineer*, Vol. 45, No. 12 (December 1975), pp. 31–33.
13. D. Gillette, "How Regulations Encourage and Discourage Innovation," *Research Management*, Vol. XX, No. 2 (March 1977), pp. 18–21.
14. A. Altshuler, "The Politics of Urban Transportation Innovation,' *Technology Review*, Vol. 79, No. 6 (May 1977), pp. 50–58.

Chapter 17

1. B. H. Klein, "The Decision-Making Problem in Development," in: National Bureau of Economic Research, *The Rate and Direction of Inventive Activity* (Princeton, N.J.: Princeton University Press, 1962), pp. 477–497; A. W. Marshall and W. H. Meckling, "Predictability of the Cost, Time and Success of Development," Ibid., pp. 461–475; K. P. Norris, "The Accuracy of Project Cost and Duration Estimates in Industrial R & D," *R & D Management*, Vol. 2 (1971), pp. 25–36; J. E. Schnee, "Research and Technological Change in the Ethical Pharmaceutical Industry," Ph.D. Diss., Univ. of Pennsylvania, Pittsburgh, 1970; R. Summers, in T. A. Marschak, T. K. Glennan, Jr., and R. Summers, *Strategy for R & D* (New York: Springer, 1967); H. Thomas, "Some Evidence on the Accuracy of Forecasts in R & D Projects," *R & D Management*, Vol. 1 (1971), pp. 55–70.

2. S. I. Doctors, *The NASA Technology Transfer Program, An Evaluation of the Dissemination System* (New York: Praeger, 1971); A. C. Cooper, "Spin-Offs and Technical Entrepreneurship," *IEEE Transactions on Engineering Management*, Vol. EM-18 (1971), pp. 2–6; A. K. Chakrabarti and A. H. Rubenstein, "Interorganiaztional Transfer of Technology," *IEEE Transactions on Engineering Management*, Vol. EM-23 (1976), pp. 20–34; L. M. Lamont, "Entrepreneurship, Technology and the University," *R & D Management*, Vol. 2 (1972), pp. 119–123; E. B. Roberts, "A Basic Study of Innovators: How to Keep and Capitalize on Their Talents," *Research Management*, Vol. XI (1968), 249 et seq.; E. B. Roberts, H. A. Wainer, "New Enterprises on Route 128,' *Science Journal*, Vol. 4 (December 1968), pp. 78 et seq.; H. Olken, "Spinn-Offs, II," *California Management Review*, Vol. 9 (1967), pp. 17–25.

3. To give two examples from a fiel that is of continued interest in this paper: P. V. Norden, "Manpower Utilization Patterns in Research and Development," Ph.D. Diss., Columbia University, 1964; IBM Technical Report 00.1191, Poughkeepsie, N.Y.; J. E. Schnee, "Research and Technological Change in the Ethical Pharmaceutical Industry," Ph.D. Diss., University of Pennsylvania, Pittsburgh, 1970.

4. Let me note in passing the lack of comparative studies of technological policies. A noted exception is A. Gerstenfeld, *Innovation: A Study of Technological Policy* (Washington, D.C.: University Press of America, 1976).

5. For this discussion see: G. Kirsch. *Systemanalytische Grundlagen der Forschungspolitik*, (Dusseldorf: Bertelsmann, 1972); G. Kirsch, Wis-

senschaft zwischen Spiel und Verpflichtung. Zur Theorie der Forschungspolitik (Freiburg/Schweiz: Universitatsverlag, 1975).
6. See J. F. Reuter, "Zur forschungspolitischen Konzeption der Bundesregierung," Schmollers Jahrbuch fur Wirtschafts- und Sozialwissenschaften, Vol. 88 (1968), pp. 51–74; G. Mensch, "Gemischtwirtschaftliche Innovationspraxis—Alternative Organisationsformen der staatlichen Forschungs und Technologiepolitik (Göttingen: Vandenhoeck und Ruprecht, 1976), pp. 46 et seq.
7. Bundesbericht Forschung V, Bundesminister for Forschung und Technologie, Ed., (Bonn, 1975).
8. These and a lot of additional policies have been discussed in depth by G. Mensch, Gemischtwirtschaftliche Innovationspraxis.
9. Bundesbericht Forschung V, p. 15, translated by the author.
10. Ibid.
11. Deutsche Wagnisfinanzierungs-Gesellschaft, Prospectus (Frankfurt, 1976), pp. 3, 10. In a talk given at the conference at which this paper was read, A. Stoehr, a board member of the company, listed possible fields of support: helping in key decisions, asking for regular review of plans, supporting marketing and. financial management functions, shaping future technical development. It remains to be seen, however, whether these objectives can be accomplished.
12. Se Technologie und Wirtschaft, Bundesminister für Wirtschaft, Ed. (Bonn, 1970), pp. 111 et seq.
13. Bundesrepublik Deutschland, Der Bundesminister fur Forschung und Technologie, Bewirtschaftungsgrundsätze für Zuwendungen auf Kostenbasis an Unternehmen der gewerblichen Wirtschaft fur Forschungs- und Entwicklungsvorhaben, i.d.F. v. 15.10.1976. (Budgeting ground rules for supporting research and development in industrial firms on a cost basis.) These guidelines are based on the "Leitsätze für die Preisermittlung auf Grund von Selbstkosten, Anlage zur Verordnung über die Preise bei öffentlichen Aufträgen vom 21.11.1953."
14. This goes far beyond standards that are discussed for other purposes. See ASC (Accounting Standards Committee), Exposure Draft 17: Accounting for Research and Development, rev. (London: April 8, 1976).
15. See my case study of a quite comparable situation: "Die Realisierung eines Zielsystems—dargestellt an Hand eines Falles," Zeitschrift für Organisation, Vol. 42 (1973), pp. 319–324, especially p. 323.
16. D. T. Campbell, "Reforms as Experiments," American Psychologist, Vol. 24 (1969), pp. 409–429 (revised version in Readings in Evaluation Research, F. G. Caro (Ed.) (New York: Russel Sage Foundation, 1971), pp. 233–261. D. T. Campbell and J. C. Stanley, Experimental and Quais-Experimental Design for Research (Chicago: Rand McNally, 1973) contains a less developed version than the first citation in this footnote.
17. See D. T. Campbell and I. C. Stanley, Experimental and Quasi-Experimental Design, p .8.
18. Ibid., p. 34.
19. Ibid., p. 57.
20. See the "BKFT 75."
21. See K. Brockhoff, "Technischer Fortschritt im Betrieb," Handwörterbuch der Wirtschaftswissenschaften, (Stuttgard: Fischer, 1977), to appear.

22. W. J. Abernathy and R. S. Rosenbloom, "Parallel and Sequential R & D Strategies," *Management Science*, Vol. 15 (1969), pp. B486–B505.
23. K. Brockhoff, *Forschungsprojekte und Forschungsprogramme. Ihre Bewertung und Auswahl* (Wiesbaden: Gabler, 1973), pp. 123 et seq.
24. K. Brockhoff, *Technischer Fortschritt im Betrieb.*
25. N. Dodson, "A General Approach to Measurement of the State of the Art and Technological Advance," *Technological Forecasting*, Vol. 1 (1970), pp. 391–408; E. N. Dodson, "Resource Analysis for R & D Programs," *IEEE Transactions on Engineering Management*, Vol. EM-19 (1972), pp. 78 et seq.
26. This was suggested elsewhere. See my paper: "Vorhersagen uber den technischen Fortschritt und Entwicklungsaufwendungen," *Zeitschrift für Betriebswirtschaft*, Vol. 43 (1973), pp. 761–776. English summary in: *German Economic Review*, Vol. 12 (1974), p. 62.
27. J. S. Mill, *A System of Logic, Ratiocinative and Inductive. Being a Connected View of the Principles of Evidence and the Methods of Scientific Investigation*, 8th ed., J. M. Robson (Ed.) (Toronto: University of Toronto Press, 1973), esp. pp. 439, 447 et seq., 449 et seq.

Chapter 18

I. C. Freeman, "The Kondrative Long Waves, Technical Change, and Unemployment," OECD (unpublished), Paris, 1977.

Chapter 22

I. W. Fellner, "Profit Maximization, Utility Maximization, and the Rate and Direction of Innovation," *The American Economic Review*, Vol. LVI, No. 2 (May 1966), pp. 24–32.

Chapter 23

1. E. Ginzberg, "The Job Problem," *Scientific American*, Vol. 237, No. 5 (November 1977), pp. 43–51.

Name Index

Abernathy, W. J., 270
Altshuler, A., 268
Ancker-Johnson, B., 260
Arrow, K., 17, 258
Ashford, N., 268
Aucoin, P., 263

Barker, William O., 75
Bean, A., 265
Beardsley, G., 258
Boffey, P. M., 260
Bradbury, F. R., 260
Brainard, Robert, v, vii
Brockhoff, K., v, 186, 269, 270
Bruno, A. V., 2, 262
Burke-Smith, P., 267

Campbell, D. T., 269
Cardon, S., 104
Caro, F. G., 269
Catling, H., 258
Chakrabarti, A. K., 266, 268
Charpie, R. A., 262
Chenea, P. F., 260, 267
Colombo, Umberto, v, 209
Colton, R. M., 261
Cooper, A. C., 104, 262, 268
Cummins, E., 258

Doctors, S. I., 268
Dodson, E. N., 270
Dolland, S. , v, 111
Douds, C. F., 266

Echterhoff-Severitt, H., 261
Ettlie, J., 266, 267
Evans, W. G., 264

Fellner, I. W., 19, 243, 270
Fels, G., 261
Fisher, W. H., 260
Freeman, C., 121, 222, 263, 265, 270
French, R., 263
Frozzo, C., 266
Fusfeld, Herbert I., v, 234

Galbraith, John K., 7
Gansler, J. S., 105, 262
Garvey, J., 266
Gergman, Mark, 266
Gerstenfeld, A., v, 176, 267, 268
Geschka, H., 266, 267
Gibbons, M., v, 121, 263, 264
Gillette, D., 73, 260, 268
Gilpin, R., 265
Ginzberg, E., 270
Glennan, T. K., Jr., 268
Griliches, Z., 17, 258
Gruber, W., 265
Gummett, P. J., 121, 263, 264
Gyllenhammar, P. G., 64

Heaton, G., 268
Hess, Earl H., v, 111
Hill, C., 266
Horn, E. J., 261
Hurter, A. P., 266
Husic, F., 258

Inose, Hiroshi, v, 140

Jevons, F. R., 264
Johnston, R., 264

Kagan, Aubrey, v, 57
Kamien, M. I., 261, 266
Kavesh, Robert A., 267
Kawase, T., 266, 267
Kelly, R., v, 41, 259
Kirsch, G., 268
Kitti, C., 266
Klein, B. H., 268

Lamont, L. M., 268
Langen, W., 261
Langrish, J., 264, 265
Lederman, L., v, 105, 159, 261, 267
Lewis, J. D., 73, 260

Maher, M., 263
Malthus, Thomas, 7, 15
Mansfield, E., v, 16, 69, 258, 259, 265
Marquis, D., 265
Marschak, T. A., 268
Marshall, A. W., 268
Martinich, J., 266
Marx, Karl, 7, 15
Matthews, R., 258
McFarlane, J., 263
Meckling, W. H., 268
Mensch, G., vi, 242, 269
Mill, J. S., 270
Miller, J. P., 266
Mogee, M., 265
Morse, R. S., 107, 262

Nason, Howard K., vi, 69
Nelson, R. R., vi, 196, 264
Noll, R. C., 265
Norden, P. V., 268
Nordhaus, W., 136, 264
Norris, K. P., 268

O'Keefe, R. D., 266

Olken, H., 268
Orths, K., 261

Pannenborg, A. E., vi, 227
Parkinson, C. N., 262
Pavitt, K., vi, 7, 223, 263
Payne, P. R., vi, 104, 262, 263
Pearson, A., 267
Peters-Koehler, B., 267

Rabinow, J., 262, 263
Radnor, M., 266
Ramo, Simon, 75
Rapoport, I., 258, 265
Reuter, J. F., 269
Roberts, E. B., 268
Robson, J. M., 270
Romeo, A., 258
Rosenbloom, R. S., 270
Rothlingshofer, Ch., 261
Rothwell, R., vi, 20, 258
Rubenstein, A. H., vi, 165, 266, 267, 268

St. Paul, R., 267
Schiffel, D., 265
Schlie, T. W., 266
Schumpeter, Joseph, 7, 136
Schnee, J. E., 268
Schwartz, N. L., 261, 266
Schwerdtner, H., 267
Shapiro, I. S., 260
Smith, Adam, 7, 65
Stanley, J. C., 269
Stoehr, A., vi, 269
Stroetmann, vi, K. A., 93, 261
Summers, R., 268

Thayer, H. E., 267
Thomas, H., 268
Thompson, C. W. N., 266
Tobin, J., 136, 264
Trozzo, C., 266

Udell, G. G., 261

Vernon, Raymond, 41, 55
Villani, E., 258

Wagner, S., 258
Wainer, H. A., 268
Walker, W., 262
Watkins, D., 267

Weidenbaum, M. L., 260, 268
Williams, B., 258
Winter, S. G., 264
Wortzel, L., 267

Zaleznik, A., 114, 263
Zegveld, Walter, vi

Subject Index

Aerospace, 144
Antitrust, 163
Attitudinal Factors, 83, 85, 105
Automobiles, 182
  electric, 148
  exhaust gases, 230
  Japan, 141
  safety, 182

Basic research, 121, 143
Belgium, 10
Bell Telephone Laboratories, 75

Common Market, 11
Communication, 141
  telecommunication, 149
Competition, 81, 83
Computers, 11, 43, 104, 106, 123,
    126, 129, 150, 152
  data processing, 94
  microprocessor, 94
  Japanese, 148, 150
  on-line, 151
  aided, 153
Council of Economic Advisors, 196
Cultural factors, 82, 83, 85, 105
Czechoslovakia, 21, 24

Dacron polyester fiber, 113
DDT, 113
Department of Agriculture, U.S., 114
Department of Commerce, U.S., 116
Department of Defense (DOD), 105,
    109

Department of Industry, U.K., 122,
    123, 126, 127
Department of Transportation, 196
Diffusion, 8, 9, 11, 93, 159
Drug Innovations, 180
duPont, 210

EEC (European Econ. Comm.), 11,
    232, 233
Electric power, 141, 144
England, 2, 10, 11, 23, 24, 45, 80-92,
    121-138, 251
Employee Inventions Law, 92
Employment, 2, 20, 83, 246
  unemployment, 92
Energy, 13, 105, 140, 148
  nuclear, 94
  nonnuclear, 106, 123
  Research Development Association
    (ERDA), 110, 115, 117
Entrepreneur, 81, 84, 85, 86, 94, 105,
    107
Experimental design, 189, 190, 191

Finance, 86, 98, 102, 148
Food industries, 144, 161
  agriculture, 235
France, 10, 11, 45, 144, 146
Future innovations, 1, 7, 140, 155
  projects, 153

General Motors, 73, 179
Germany, 10, 11, 45, 80-92, 93, 144,
    146

Subject Index

Government procurement programs, 84
Growth, economic, 8, 80, 83, 84, 90, 93, 94, 140, 241

Health technology, 94, 95, 228
  hospital automation, 153
  medical information systems, 153

IBM, 210
Inflation, 9
Information, 96, 101, 140
  pollution, 155
  technology, 153
Infrastructure:
  scientific-technological, 95
Innovations
  effects on employment, 1, 4, 20, 65
  incremental, 22, 23
  in Japan, 155-157
  major beneficiary, 1, 16
  radical, 22, 23
Inventions, 81, 85, 141
Inventor, 81, 86
Iron & Steel, 141, 144
Italy, 10, 11, 146

Japan, 2, 10, 45, 48, 140-164, 210, 211, 251
Jet engine, 113, 148

Key technologies program, 85

Leather production, 93
Less developed countries (LDC's), 8, 56
Little, Arthur D., 80

Management, 94, 114, 153, 189, 254
  incentives, 108
  of innovation, 101
Manpower, 96, 101
Marketing, 7, 81
  marketable products, 100
  marketing of innovation, 101

markets, 95, 97
Materials development, 94
Media, 106
Microwave transmission, 149
Ministry of International Trade and Industry (MITI), 147, 148, 153
M.I.T., 183
Mobility of individuals, 84

National Research Development Corporation (NRDC), 84, 85, 86
National Science Foundation, 198, 200
Netherlands, 10, 11
Nuclear:
  energy, 94
  proliferation, 12

O.E.C.D., 7, 8, 10, 11, 42, 45, 50, 223, 234
Office of Management & Budget, 196, 197
Office of Science & Tech. Pol, 197
Oil, 141, 161
  petro-chemical industries, 141, 144
Optics, 94

Patents, 81, 82, 83, 84, 86, 163
  licensing, 100
  procurement, 165
Penicillin, 113
Pharmaceuticals, 180, 214
Polaroid camera, 113
Pollution control, 95
  information, 155
Private rate of return, 16, 18
  private sector, 159
Productivity, 8, 10, 81, 141, 199
Psychology, 242-244
  innovation, 3
  of work groups, 2
  market, 106
  stock market, 106
Public sector, 159

Recessions, Economic, 93
Regulation, 2, 78, 95, 114, 116, 162, 163, 176-185, 253
Research Applied to National Needs (RANN), 117
Risks, technological, 81, 87, 107, 108

Saab, 63
San Francisco:
  new technology based firms, 104
Ship and marine technology, 123
Small Business Administration, 101, 109, 116, 117
Small firms, 2, 81, 85, 86, 93, 94, 95, 104, 105
  high technology, 111, 113, 114, 116
  burden for social demands, 100
Social, 1, 168, 220
Social rate of return, 16, 17, 18
Social demand, 93, 100
Social change, 95, 106, 140
Social safer machinery, 100
Soviet Union, 12, 146
Space technology, 11, 94
N.A.S.A., 109, 110, 115, 200
Stagflation, 9
Standard Oil, 210
Standardization, 229
Stock market, 106
  Securities Exchange Commission, 108
  shareholders, 92
Strategy, corporate, 96, 121, 128, 131
  Japan, 141

for innovation, 2
in Europe, 216
Supersonic transport, 197
Sweden, 2, 57, 58
Switzerland, 10, 11, 214, 215

TRW, 75
Taxation, 81, 82, 83, 84, 105, 107, 108, 115, 116
Technology gap, 10, 230
Technology, intensive, 43, 46, 54
Telephones, 150, 154
  electronic switching systems, 149
Television, 141
Textile, 20, 21, 24, 26, 129, 239
Trade patterns, 41, 141, 144, 150, 155, 247
  exports, 2, 83
  imports, 93
Transportation, 141, 144, 153
  equipment, 13

United States, 2, 9, 11, 19, 45, 46, 83, 104, 146, 249

Venture, capital, 82, 84, 86, 87, 104, 105, 108
Volvo, 2, 14, 60, 248

Wages, 13
Western Electric, 210
Working conditions, 2, 95

Xerox, 215